Strategies for Combatting Multidrug-Resistant and Extensively Drug-Resistant Bacteria: Current Challenges and Future Prospects

Strategies for Combatting Multidrug-Resistant and Extensively Drug-Resistant Bacteria: Current Challenges and Future Prospects

Guest Editors

Adriána Liptáková
Bożena Futoma-Kołoch

MDPI

Basel • Beijing • Wuhan • Barcelona • Belgrade • Novi Sad • Cluj • Manchester

Guest Editors

Adriána Liptáková
Institute of Microbiology
University Hospital Bratislava
Bratislava
Slovakia

Bożena Futoma-Kołoch
Department of Microbiology
University of Wroclaw
Wroclaw
Poland

Editorial Office
MDPI AG
Grosspeteranlage 5
4052 Basel, Switzerland

This is a reprint of the Special Issue, published open access by the journal *Antibiotics* (ISSN 2079-6382), freely accessible at: https://www.mdpi.com/journal/antibiotics/special_issues/7086E5C3O5.

For citation purposes, cite each article independently as indicated on the article page online and as indicated below:

Lastname, A.A.; Lastname, B.B. Article Title. *Journal Name* **Year**, *Volume Number*, Page Range.

ISBN 978-3-7258-7625-9 (Hbk)
ISBN 978-3-7258-7626-6 (PDF)
https://doi.org/10.3390/books978-3-7258-7626-6

Contents

About the Editors

Adriána Liptáková

Adriána Liptáková, PhD., MPH, univ. prof. is the head of the Institute of Microbiology of the Faculty of Medicine of Comenius University and University Hospital Bratislava. She works as a doctor specializing in clinical microbiology and has many years of experience in organizational work, leading research teams, and implementing domestic and international grant tasks.

In her scientific and research activities, she systematically focuses on the following:

The issue of antimicrobial resistance (AMR) in important bacterial pathogens.

Monitoring resistance trends in the hospital environment.

Searching for alternative therapeutic methods in the fight against resistant strains, in particular the use of phage therapy and autovaccines.

Integration of modern diagnostic and molecular methods (e.g., next-generation sequencing, MALDI-TOF, MLST) into clinical practice.

She is the author and co-author of dozens of publications in renowned scientific journals indexed in WoS and Scopus databases, with her works mainly devoted to ESBL- and carbapenem-resistant bacteria and multiresistant strains of *Escherichia coli* and *Klebsiella pneumoniae*, as well as experimental phage therapy.

Significant outputs include a publication on phage therapy of urinary tract infections caused by the genus *Cronobacter*, which demonstrated the prospect of clinical use of phages in infectious diseases.

Bożena Futoma-Kołoch

Bożena Futoma-Kołoch is a microbiologist with extensive research and teaching experience, affiliated with the Department of Microbiology at the Faculty of Biological Sciences, University of Wrocław, Poland. *Her research demonstrates an interest in antimicrobial therapy and bacterial resistance. She has authored and co-authored multiple studies focused on antibiotic resistance mechanisms, susceptibility profiling of pathogenic bacteria such as Escherichia coli and Salmonella enterica, and the interactions of bacterial cells with antimicrobial agents. Her work encompasses resistance determinants and applied aspects of antimicrobial susceptibility and intervention strategies.* She is the author of 84 scientific publications and 37 popular science papers with a total Impact Factor of 101. Her scholarly output is reflected in an h-index of 11 and more than 900 citations recorded in the Web of Science and Scopus databases.In addition to her research activity, she is actively involved in science communication, academic teaching, and organizational work. She has taught a wide range of university courses, including microbiology, bacteriology, vaccinology, and bacterial growth and control, and has contributed to the organization of scientific conferences. Moreover, she acts as an expert for major research and funding institutions, including the Polish National Agency for Academic Exchange (NAWA) and the National Centre for Research and Development (NCBR). She is a member of three COST Actions (European Cooperation in Science and Technology): CA22166 *"Safety in the Game Meat Chain"* (2023-2027), CA23111 *"Searching for Nanostructured or Pore-Forming Peptides for Therapy"* (2024-2028), and CA24127 *"Pan European Commission on Photoantimicrobial Testing"* (2025-2029).

Preface

Antimicrobial resistance is currently considered one of the most important threats to global public health by the World Health Organization. Globally, multidrug-resistant (MDR) and extensively drug-resistant (EDR) bacterial pathogens are spreading. Although the acquisition of resistance is a natural process, the misuse of antibiotics, inadequate surveillance, and the poorly controlled regulation of antibiotics in clinical medicine and in the livestock industry potentiate the rise and worldwide spread of MDR bacteria.

Mortality caused by resistance to antibiotics is a major health problem, causing more than 33,000 deaths in Europe annually. Although this figure is already alarming, the number of deaths worldwide is predicted to increase to more than 10 million by 2050; however, predictions are difficult to make in this field.

We should focus on new strategies, including those that are not only related to novel antimicrobial drugs but also those related to uncommon drug strategies, such as the use of bacteriophages, autovaccines, or personalized microbiota, and antimicrobial stewardship with updated guidelines.

Adriána Liptáková and Bożena Futoma-Kołoch
Guest Editors

antibiotics

MDPI

rticle

High Diversity but Monodominance of Multidrug-Resistant Bacteria in Immunocompromised Pediatric Patients with Acute Lymphoblastic Leukemia Developing GVHD Are Not Associated with Changes in Gut Mycobiome

ara Sardzikova [1], Kristina Andrijkova [1], Peter Svec [2], Gabor Beke [3], Lubos Klucar [3], Gabriel Minarik [4], Viktor Bielik [5], Alexandra Kolenova [2] and Katarina Soltys [1,*]

1 Department of Microbiology and Virology, Faculty of Natural Sciences, Comenius University in Bratislava, 841 04 Bratislava, Slovakia
2 Department of Pediatric Hematology and Oncology, Children's Haematology and Oncology Clinic, Faculty of Medicine, Comenius University in Bratislava, 833 40 Bratislava, Slovakia
3 Institute of Molecular Biology, Slovak Academy of Sciences, 841 04 Bratislava, Slovakia
4 Medirex Group Academy n.p.o., Novozamocka 67, 949 05 Nitra, Slovakia
5 Department of Biological and Medical Science, Faculty of Physical Education and Sport, Comenius University in Bratislava, 814 69 Bratislava, Slovakia
* Correspondence: katarina.soltys@gmail.com

Abstract: Graft-versus-host disease (GvHD) is a severe complication after hematopoietic stem cell transplantation (HSCT). Our study focused on identifying multidrug-resistant (MDR) gut bacteria associated with GvHD-prone guts and association with gut microbiota (GM) diversity, bacteriome, and mycobiome composition in post-HSCT patients. We examined 11 pediatric patients with acute lymphoblastic leukemia (ALL), including six with GvHD, within three time points: seven days pre-HSCT, seven days post-, and 28 days post-HSCT. The gut microbiome and its resistome were investigated using metagenomic sequencing, taxonomically classified with Kraken2, and statistically evaluated for significance using appropriate tests. We observed an increase in the abundance of MDR bacteria, mainly *Enterococcus faecium* strains carrying *msr(C)*, *erm(T)*, *aac(6′)-li*, *dfrG*, and *ant(6)-la* genes, in GvHD patients one week post-HSCT. Conversely, non-GvHD patients had more MDR beneficial bacteria pre-HSCT, promoting immunosurveillance, with resistance genes increasing one-month post-HSCT. MDR beneficial bacteria included the anti-inflammatory *Bacteroides fragilis*, *Ruminococcus gnavus*, and *Turicibacter*, while most MDR bacteria represented the dominant species of GM. Changes in the gut mycobiome were not associated with MDR bacterial monodominance or GvHD. Significant α-diversity decline (Shannon index) one week and one month post-HSCT in GvHD patients ($p < 0.05$) was accompanied by increased *Pseudomonadota* and decreased *Bacteroidota* post-HSCT. Our findings suggest that MDR commensal gut bacteria may preserve diversity and enhance immunosurveillance, potentially preventing GvHD in pediatric ALL patients undergoing HSCT. This observation has therapeutic implications.

Keywords: multidrug resistance; gut microbiome; gut mycobiome

1. Introduction

The human intestine possesses a complex ecosystem of trillions of microorganisms, i.e., the gut microbiota, which can turn into a place with specific pathophysiological conditions triggering the human immune system during hematopoietic stem cell transplantation (HSCT). One such condition, graft-versus-host-disease (GvHD), is a life-threatening complication of HSCT. During the development of GvHD, immunologically competent cells, especially donor T cells, are activated, leading to a response to recipient tissue antigens, and subsequently, human organs, including the hematopoietic system, the liver, and the intestine. The different histocompatibility complexes of donor and recipient, pretransplant chemotherapy, and gut radiation damage contribute to GvHD [1].

It is already an accepted fact that the intestinal microbiota has a significant impact on GvHD development. Studies on germ-free (GF) mice have shown a reduction in the risk of GvHD development in GF mice [2–4]. However, although progress in therapeutic advancements has significantly improved HSCT outcomes, multiresistant bacteria remain a challenge. The gut microbiota represents a reservoir of antimicrobial resistance genes. While imbalances in the gut microbiota are tightly associated with therapy and conditioning regimen, as well as parenteral diet, immune suppression leads to the colonization of resistant bacteria, often identified as the cause of bloodstream infections or sepsis [5].

Multiresistant bacteria, often referred to as multidrug-resistant (MDR) bacteria, are strains which have developed resistance to multiple antibiotics, limiting treatment options and increasing the risk of severe infections. Although the gut microbiota has been intensively studied in the context of HSCT treatment, studies investigating the gut resistome are lacking. Although some studies have provided information on the association of MDR colonization with increased bloodstream infection [6], others have not shown a significant impact on transplantation outcome [7]. Furthermore, studies on the loss of diversity and the expansion of resistant bacteria which colonize depleted environments and become dominant species in the gut are scarce.

The enrichment of *Enterococcaceae*, *Staphylococcaceae*, *Streptococcaceae*, and *Pseudomonadota* was previously correlated with GvHD development. Moreover, *Prevotella* spp. or *Escherichia coli* are risk factors for GvHD [8]. In addition, GvHD is associated with deficiencies of *Lactobacillus johnsonii*, *Clostridium leptum*, and *Eubacterium rectale* in pediatric patients and with a lack of *Bacteroides* and *Parabacteroides* genera in adult patients [9,10]. GvHD development has also been associated with a low abundance of *Blautia* and *Akkermansia muciniphila* early post-HSCT [11].

Although the role of the gut mycobiome has been stressed in a few studies, to the best of our knowledge, there is no published evidence on its composition in pediatric oncology patients undergoing HSCT. Furthermore, no existing study has associated the dominance of multidrug-resistant bacteria with gut mycobiome composition and the implications for clinical outcomes. Understanding the dynamics of multiresistant bacteria in the context of GvHD is essential for the development of effective prevention and treatment strategies.

We hypothesize that GvHD patients will demonstrate a unique microbial composition, characterized by lower diversity and increased abundance of resistant microorganisms within the gut microbiome. In this study, we aimed to analyze the gut bacteriome and mycobiome composition and the "gut profile" of resistant microorganisms. We focused on the association of the gut bacteriome and mycobiome and multidrug-resistant bacteria dominance within the gut of pediatric oncology patients with GvHD outcome after allo-HSCT. Finally, we investigated the diagnostic potential of the gut microbiome and resistome for GvHD development.

2. Results

2.1. Subject Characterization

We made a deep gut microbiome analysis of pediatric oncology patients (32) based on the therapy, age, BMI, sex, and diagnosis and determined the associations between established parameters and GvHD development. We did not confirm any association of these parameters with GvHD, except for myelodysplastic syndrome diagnosis, where all six patients were found to be negative for GvHD development (Pearson chi-square $p = 0.025$) (Figure 1). Considering the intragroup variability regarding the GvHD+/GvHD− group of patients, several bacterial genera were differentially abundant.

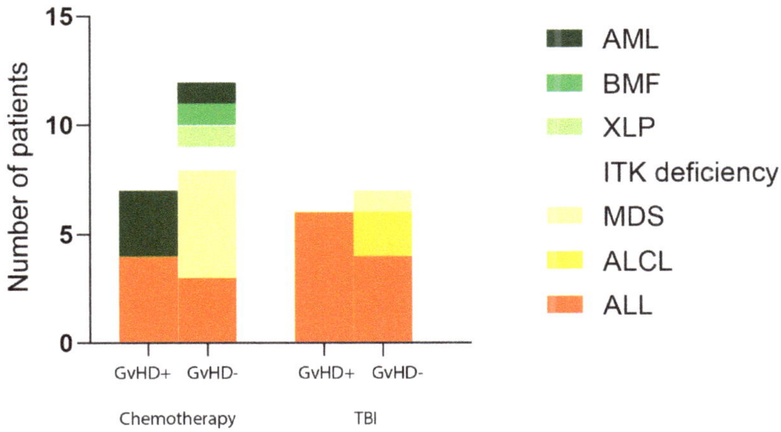

Figure 1. Graphical visualization of patients according to chemotherapy or TBI; groups with or without GvHD development. Different colors represent individual diagnoses.

Before HSCT, the gut mucosa layers associated with *Actinomyces* (p = 0.024) and *Rumunicoccus torques* (p = 0.039) were found to be more abundant in females compared to males, suggesting that females could suffer more from a thinner mucus layer, caused by mucus-degrading *Ruminococcus torques;* in these cases, the development of actinomycosis is usually associated with the break-down of the mucosal barrier. Beneficial, short-chain fatty acid (SCFA)-producing bacterial genera *Dorea* and *Eubacterium hallii* prevailed in the guts of patients with normal BMI (18.5–24.9) compared to low BMI (<18.5) (p = 0.012). Regarding the age of patients, the significantly lower relative abundance of *Enterococcus* (p = 0.034) and higher relative abundance of *Peptostreptococcaceae* (p = 0.025) in the guts of patients older than ten years was determined.

One week after HSCT, *Streptococcus* (p = 0.006) abundance was typical for patients with low BMI, while the abundances of *Abiotrophia* (p = 0.037) and *Veillonella* (p = 0.041) were significant for patients younger than ten years of age.

After one month from HSCT, the *Staphylococcaceae* family prevailed in the gut of patients older than ten years (p = 0.033). There were no complex changes or differences between patients based on diagnosis, therapy, BMI, sex, or age. However, several opportune pathogen bacterial genera, like *Enterococcus* (younger than ten years), *Actinomyces* (female), *Streptococcus* (low BMI), or *Staphylococcaceae* (older than ten years), were found to dominate in the gut of patients in certain stages of therapy related to HSCT.

The analysis of anthropometric parameters and clinical diagnoses of all included patients did not reveal any discernible pattern for the development of GvHD, suggesting that although all patients have undergone a conditioning regimen before HSCT, harsh treatment did not unite the gut microbiome, resulting in a pattern typical for GvHD development. Since the number of patients suffering from the same oncology diagnoses was low, except for ALL, we decided to focus our further analysis exclusively on patients diagnosed with ALL. This decision was guided by the need for a more homogenous patient group to draw meaningful conclusions and to ensure that our study's outcomes would be as relevant and specific as possible to this patient population.

2.2. Multiresistant Bacteria in the Gut of Patients with ALL

The detection of multiresistant bacteria within the gut microbiome has become increasingly critical for our understanding of their potential impact on human health. Multiresistant bacteria are defined as bacteria with two or more resistance genes within their genomes. In this study, resistance genes localized either on chromosomes or on plasmids were included.

Gut resistome analyses uncovered intriguing trends among patients with and without GvHD. Notably, we observed a significant increase in multiresistant bacteria one week after HSCT in patients afflicted by GvHD. At the same time, patients without GvHD predominantly hosted known commensal bacteria with multidrug resistances in the initial phase of the HSCT (before HSCT). Strikingly, one month after HSCT, we observed a rise in resistance genes among patients without GvHD, suggesting a window of opportunity for predicting infection risks. Moreover, our comprehensive analyses of gut microbiota composition unveiled the dominance of multiresistant bacteria, further emphasizing their role as critical actors in gut shaping. The gut of GvHD patients was primarily inhabited by multiresistant bacteria *E. faecium*, i.e., mainly *E. faecium* ME3, *E. faecium* ICU-1-2, *E. faecium* KUHS13, and *E. faecium* VVEswe-R. These microorganisms most often possessed a combination of the *msr(C)*, erm(T), *aac(6′)-li*, *dfrG*, and *ant(6)-la* genes (Table 1). A combination of resistance genes was shown in the second most commonly identified multiresistant bacterium, *Klebsiella pneumoniae*, carrying a combination of the *aph(3″)-Ib*, *sul2*, and *aph(6)-ld* genes (Table 1).

Table 1. List of multidrug-resistant bacteria in the gut of patients undergoing HSCT.

Patient	d − 7		d + 7		d + 28	
			GvHD PATIENTS			
P2	*Klebsiella pneumoniae* strain 314013 plasmid p314013-KPC	*blaTEM-181*, *blaTEM-1B*, *blaTEM-1C*	*Enterococcus faecium* KUHS13 DNA	*dfrG*, *msr(C)*, *ant(6)-Ia*	*Enterococcus faecium* KUHS13 DNA	*dfrG*, *msr(C)*, *ant(6)-Ia*
	Salmonella enterica subsp. enterica serovar Manhattan strain SA20084699 plasmid unnamed2	*aph(3″)-Ib*, *aph(6)-Id*, *dfrA14*, *sul2*	*Enterococcus faecium* strain ME3	*aac(6′)-Ii*, *erm(T)*	*Enterococcus faecium* strain ME3	*aac(6′)-Ii*, *erm(T)*
P3	*Klebsiella pneumoniae* subsp. pneumoniae strain 21 plasmid pKpQIL_pKpN	*blaTEM-1B*, *aph(3″)-Ib*, *sul2*	*Staphylococcus haemolyticus* strain CGMH-SH51	*dfrG*, *mecA*, *tet(K)*	*Staphylococcus haemolyticus* strain Sh29/312/L2	*mef(A)*, *msr(D)*
P5			*Staphylococcus haemolyticus* strain Sh29/312/L2	*mef(A)*, *mph(C)*, *msr(A)*, *msr(D)*, *qacA*, *aph(3′)-III*, *blaZ*		
			Enterococcus faecium strain ME3	*aac(6′)-Ii*, *dfrG*, *msr(C)*, *aac(6′)-aph(2″)*	*Enterococcus faecium* strain ME3	*aac(6′)-Ii*, *dfrG*, *msr(C)*
			Enterococcus faecium strain VVEswe-R plasmid pVVEswe-R4	*ant(6)-Ia*, *aph(3′)-III*, *erm(B)*	*Enterococcus faecium* strain VVEswe-R plasmid pVVEswe-R4	*ant(6)-Ia*, *aph(3′)-III*, *erm(B)*
P6	*Enterococcus faecium* KUHS13 DNA	*msr(C)*, *ant(6)-Ia*, *aac(6′)-Ii*, *dfrG*	*Enterococcus faecium* KUHS13 DNA	*msr(C)*, *ant(6)-Ia*, *dfrG*, *aac(6′)-Ii*		
P8	*[Ruminococcus] gnavus* strain NBRC 114413	*catP*, *tet(O)*			*Enterococcus faecium* strain ICU-2-1	*aac(6′)-Ii*, *VanHAX*, *dfrG*

Table 1. *Cont.*

Patient	d − 7		d + 7		d + 28	
P10	*Enterobacter cloaceae* strain EclC2185 plasmid pEclC2185CTXM15	*aph(3″)-Ib, aph(6)-Id, blaCTX-M-15, blaTEM-1B, sul2, aac(3)-IIa, aac(6′)-Ib-cr, blaOXA-1*	*Clostridioides difficile* strain DSM 104450	*ant(6)-la, erm(B)*	*Enterococcus faecium* strain VVEswe-R	*msr(C), ant(6)-Ia, aph(3′)-III*
					Enterobacter cloaceae strain EclC2185 plasmid pEclC2185CTXM15	*aph(3″)-Ib, aph(6)-Id, blaCTX-M-15, blaTEM-1B, sul2, aac(3)-IIa, aac(6′)-Ib-cr, blaOXA-1*
			non-GvHD PATIENTS			
P1	*Enterococcus faecium* strain HY07	*acc(6′)-Ii, tet(X)*			*Enterobacter hormaechei* subsp. steigerwaltii strain 34998 plasmid p34998-210.894kb	*aph(3″)-Ib, aph(6)-Id, dfrA14, sul2, blaACT-15*
	Bacteroides fragilis strain FDAARGOS	*tet(X), cfxA5*			*Enterococcus faecium* strain ICU-1-2	*ant(6)-Ia, aph(3′)-III*
					Enterococcus faecium strain VB12993 plasmid pVB12993_p3	*aph(3′)-III, acc(6′)-aph(2″), msr(C)*
P4	*Enterococcus faecalis* isolate WE0851	*erm(B), dfrG, lsa(A)*				
	Bacteroides fragilis strain BFG-569	*cfxA, cfxA3, cfxA4, cfxA5*				
	Enterococcus faecium isolate E8440 plasmid: 4	*aac(6′)-Ii, ant(6)-Ia, aph(3′)-III, msr(C)*				
	[Ruminococcus] gnavus strain NBRC 114413	*tet(O), tet(W)*				
	Staphylococcus aureus strain NX-T55	*lnu(B), lsa(E)*				
P7	*Klebsiella pneumoniae* strain QD23	*aph(3″)-Ib, aph(6)-Id, sul2*			Enterococcus faecium KUHS13 DNA	*dfrG, msr(C), ant(6)-Ia*
	Enterocloster clostridioformis FDAARGOS_739	*tet(40), tet(O)*				
	Turicibacter sp. H121	*mef(A), msr(D)*				

Table 1. *Cont.*

Patient	d − 7		d + 7		d + 28	
P9	*Enterococcus faecium* strain VVEswe-R	*msr(C), aph(3′)-III, aac(6′)-aph(2″), erm(B)*			*Enterococcus faecium* strain VVEswe-R	*msr(C), aac(6′)-aph(2″), erm(B)*
	Enterococcus faecium strain ICU-1-2	*aac(6′)-Ii, erm(T)*				
P11					*Staphylococcus haemolyticus* strain CGMH-SH51	*aac(6′)-aph(2″), dfrG, mecA, mph(C), msr(A)*

Our investigation also revealed multiresistant commensals *B. fragilis* (two patients), *Rumincoccus gnavus* (one patient), *Enterocloster clostridioformis* (one patient), or *Turicibacter* (one patient) in the gut of non-GvHD patients before HSCT (Table 1). Remarkably, one of the GvHD patients benefited from multiresistant commensal *bacteria [Ruminococcus] gnavus*, while others showed the presence of opportunistic pathogens. Nevertheless, non-GvHD patients also had multiresistant bacteria such as *E. faecium* (three patients), *K. pneumoniae* (one patient), or *S. aureus* (one patient) before HSCT. Still, their presence was no longer confirmed one week after HSCT (Table 1). On the contrary, the GvHD patients showed the existence of these bacteria mostly one week after HSCT. *S. haemolyticus* strains with resistance genes *erm(C), dfrG, mecA, tet(K), mef(A), mph(C), msr(A), msr(D), qacA, aph(3′)-III*, and *blaZ* was detected in one patient's gut microbiome. *E. faecium* strains were identified in three other patients, and the other one had *Clostridioides difficile* in his gut microbiome (Table 1).

It is important to note that some of the genes included in this study are multidrug resistance (MDR) genes, i.e., exhibiting resistance to two or more antimicrobials. The *aac(6′)-li* gene is responsible for resistance to dibekacin, gentamicin, netilmicin, or tobramycin from the aminoglycoside family. The *bla-ACT-15* gene confers resistance to amoxicillin, ampicillin, cefotaxime, cefoxitin, and ceftazidime from the beta-lactam family. The *msr(C)* gene provides resistance to erythromycin and telithromycin from macrolides, as well as pristinamycin IA, quinupristin, and virginiamycin S from the streptogramin B class of antimicrobials. Figure 2 shows the specific antimicrobials and antimicrobial classes associated with each identified gene.

2.3. Antimicrobials

Of the twenty observed groups of antimicrobials, the most prevalent one in the gut microbiome of 67% of GvHD patients before HSCT, 100% of patients one week after HSCT and one month after HSCT, 80% of non-GvHD patients before HSCT, 20% of patients one week after HSCT, and 40% of patients one month after HSCT was aminoglycoside. Furthermore, another three groups of antimicrobials were highly abundant: lincosamides, folate pathway antagonist, macrolides, and beta-lactams. Lincosamide resistance genes were identified in 50% of GvHD patients' gut microbiomes before HSCT, 50% of patients after one week, and 84% of patients one month after HSCT. The resistance genes from this group were also detected in the gut microbiome of 40% of patients without GvHD before HSCT, 80% of patients after one week, and 60% one month after HSCT. Folate pathway antagonist was detected in 84% of GvHD patients' gut microbiomes before and one week after HSCT, while it was only detected in 67% of patients one month after HSCT. Its presence was detected in the gut microbiome of 40% of non-GvHD patients before HSCT and three patients one month after HSCT. Macrolide resistance genes were found in the gut microbiome of 34% of GvHD patients and 40% of non-GvHD patients before and 84% of GvHD and 40% of non-GvHD patients one week after HSCT, and finally, in 67% of GvHD patients and 100% of non-GvHD patients one month after HSCT. Tetracyclines were detected in 84% of GvHD patients' gut microbiomes before HSCT and 60% of non-GVHD patients, while its presence decreased to 34% of GvHD and 40% of non-GvHD patients one week after HSCT and 17% of GvHD patient one month after HSCT. Streptogramin b resistance genes were detected in the gut microbiome of (GvHD/non-GvHD) 34%/40% patients before HSCT, 84%/40% patients one week after HSCT, and 84%/80% one month after HSCT. Before HSCT and one week after HSCT, five patients, 50% of the GvHD, and 40% of the non-GvHD group were potentially resistant to beta-lactams, as were 50% of GvHD patients together with 40% of non-GvHD patients one month after HSCT (Figure 3). The chi-square test showed a significantly positive association between beta-lactam resistance and GvHD outcome ($p = 0.041$). Of the 14 healthy controls, only eight carried resistance genes primarily for tetracycline (*tet(Q)*, *tet(O)*, *tet(W)*), while three controls carried resistance genes for also beta-lactams (*cfxA*). Folate pathway antagonist (*sul2*) was found in one healthy individual (Figure 3).

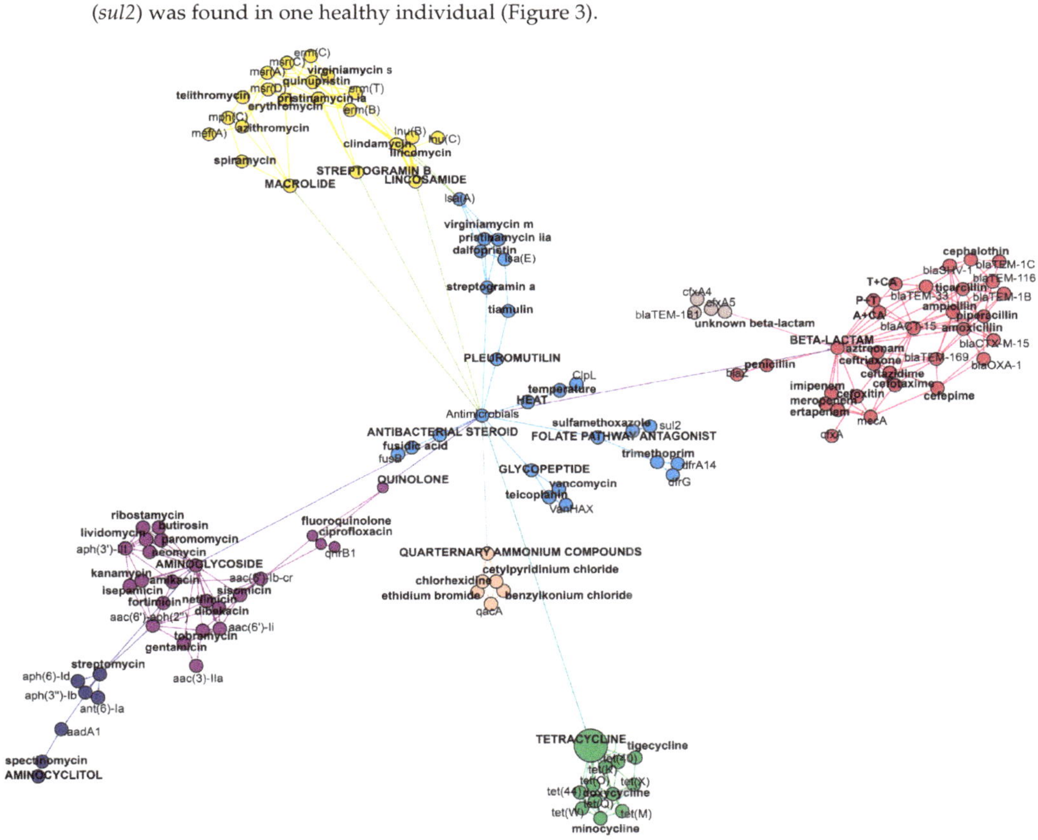

Figure 2. This network shows the relationship between individual resistance genes and antimicrobials. Different colors represent different antimicrobial classes (aminoglycoside—purple; tetracycline—green, macrolide, streptogramin b and, lincosamide—yellow; beta-lactam—pink; quarternary ammonium compounds—beige, unknown beta-lactam—brown), while light blue was used for the class of antimicrobials that were represented with a low number of resistance genes and antimicrobials (glycopeptide, antibacterial steroid, pleuromutilin, folate pathway antagonist, aminocyclitol). P + T—piperacillin + tazobactam. A + CA—ampicillin + clavulanic acid. T + CA—ticarcillin + clavulanic acid.

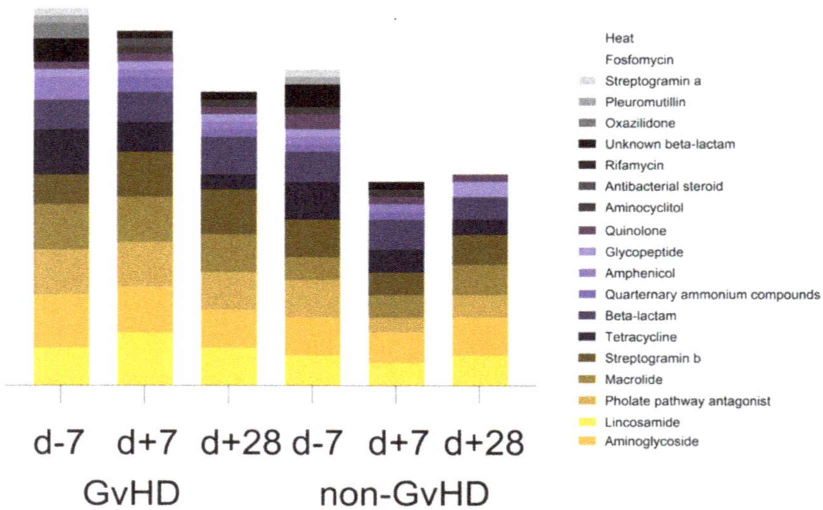

Figure 3. Groups of antimicrobials represented by antimicrobial resistance genes detected in the gut of pediatric oncology patients.

2.4. Resistance and Virulence Genes of the Gut Microbiota of Pediatric Oncology Patients Undergoing HSCT

The ResFinder identified 85 resistance genes across the samples. The resistance gene *aac(6′)-Ii* (aminoglycoside) was the most frequently found gene in the following patient groups: 60% of patients without GvHD and 17% of with GvHD before HSCT, 50% of GvHD and 40% of non-GvHD patients one week after HSCT, and 40% of patients without GvHD and 50% of patients with GvHD one month after HSCT.

In GvHD patients before HSCT, *sul2* was the most frequently detected resistance gene in the gut microbiome, found in 67% of patients, compared to only 20% of non-GvHD patients. The *tet(W)* gene was also detected in the same percentage of patients before HSCT. *aph(3′′′)-lb* was identified in three GvHD and one non-GvHD patient before HSCT, while *blaTEM-1B* was found only in three GvHD patients before HSCT. *dfrg* was identified only in 67% of GvHD patients one week after HSCT in the gut microbiome, while *erm(T)* was detected in only 60% of non-GvHD patients.

We found that antibiotic resistance genes *ant(6)-la* and *msr(C)* were present in the gut microbiome of 50% of patients with GvHD, and *msr(C)* was detected in 40% of non-GvHD patients, while *ant(6)-la* was found in only 20% of non-GvHD patients. After a month of HSCT, a similar number of resistance genes were detected in all patient samples. *msr(C)* and *ant(6)-la* were found in both groups of patients, while *aph(3)-III* with *erm(B)* was detected in 33% of patients with GvHD and 40% of patients without GvHD. The *dfrG* gene was present in 50% of patients with GvHD and 40% without GvHD; it was significantly (p = 0.019) more often detected in GvHD patients. The erm(T) gene showed a trend (p = 0.074) of being more common in patients with GvHD.

The patients who did not develop GvHD were found to be positive for *Enterococcus faecalis* isolates identified before HSCT. Among these patients, one was positive with two isolates: *E. faecalis* 28157 and WE0851. *E. faecalis* 28157 was found to carry five virulence genes, i.e., endocarditis-specific antigen (*efaA*), sugar-binding transcriptional regulator (*LacI* family, *bopD*), endocarditis and biofilm-associated pilus tip protein (*EbpA*), as well as its major (*ebpC*) and minor (*ebpB*) subunits. On the other hand, the isolate WE0851 carried three identified virulence genes: undecaprenyl diphosphate synthase (*cpsA*), phosphatidate cytidylyltransferase (*cpsB*), and surface protein *Fss1* fibrinogen-binding protein (*fss1*). One GvHD patient carried virulence gene *pcr4*, found in the type III secretion system protein of the *Pseudomonas aeruginosa* strain PA59 one week after HSCT, and in another patient with GvHD, *E. faecium* carried the virulence gene *acm* coding for collagen adhesin precursor. The same gene was identified in 33% of GvHD patients before HSCT and 50% of patients one week after HSCT. Notably, all of these *E. faecium* strains, except for that found in one patient, also harbored the resistance gene *aac(6′)-Ii*, which was found on the chromosome.

2.5. Multidrug-Resistant Commensal Bacteria of Gut of Pediatric Patients Suffering from Acute Lymphoblastic Leukemia during HSCT

In the gut microbiome of GvHD patients, we detected several opportune pathogens and commensals that carry resistance genes in their genomes. However, *Ruminococcus gnavus* NBRC 114413 with resistance genes *catP* and *tet(O)*, *Bacteroides fragilis* DCMOUH0017B with resistance genes *cfxA*, *tet(Q)* and *qnrD1*, and *Parabacteroides distasonis* ATCC 8503 were found in non-GvHD patients. Additionally, *Lachnospiraceae* bacterium strains also carried resistance gene *tet(W)* only in GvHD patients before HSCT. In GvHD patients, we detected only one commensal beneficial bacterium and *Latilactobacillus curvatus* with *ClpL* resistance gene one week after HSCT.

On the other hand, non-GvHD patients had a higher frequency of detection of resistance genes in their commensal bacteria. *Erysipelotrichaceae* bacterium LH1062 with *tet(44)* gene was detected together with *Akkermansia muciniphila* Akk2670 (*lnu(C)*), the gut barrier protector, before HSCT. Before HSCT, *Turicibacter* sp. H121 (*mef(A)*, *msr(D)*) was also detected. Resistance genes were detected in *Bacteroides uniformis* NBRC 113350 (*tet(Q)*), *B. fragilis* (*cfxA5*), and *Streptococcus salivarius* (*tet(M)*) one week after HSCT. *Faecalibacterium* sp. IP-3-29 (*aph(3′)-III*) and *Lachnospiraceae* bacterium (*tet(W)*) were abundant bacteria carrying resistance genes one month after HSCT. Nevertheless, patients with resistant or multiresistant commensal bacterial strains or isolates often harbored opportune pathogen bacterial strains such as *E. faecium*, *S. haemolyticus*, or *K. pneumoniae*. Remarkably, only one patient who did not develop GvHD did not show resistance carried by non-pathogenic bacteria (Figure 4).

In the healthy controls, resistance genes were present exclusively in beneficial commensal bacteria, namely, *B. fragilis* with *tet(Q)* and *cfxA* genes, *Phocaeicola dorei* with *tet(Q)* gene, and *P. distanosis* with *tet(Q)* and *cfxA* genes, as well as *Cutibacterium* acnes with *tet(W)* gene.

2.6. Multidrug-Resistant Opportune Pathogens in the Gut of Pediatric Patients Suffering from Acute Lymphoblastic Leukemia during HSCT

The non-GvHD patients also had detectable *E. faecium* KUHS13 (*dfrG*, *msr(C)*, *ant(6)-la*) and VVEswe-R (*msr(C)*, *aph(3′)-III*, *erm(B)*, *aac(6′)-aph(2″)*) at various timepoints. Additionally, *K. pneumoniae* with resistance genes *aph(3″)-Ib*, *aph(6)-Id*, and *sul2* was detected in these patients prior to the HSCT procedure. Both bacterial species were detected in GvHD and non-GvHD patients before HSCT. MDR *E. faecium* was detected in 60% of non-GvHD but only in 17% of GvHD patients before HSCT and 67% of GvHD patients one week and one month after HSCT. In non-GvHD patients, these species were detected in the gut only one month after HSCT.

Specifically, patients with GvHD were identified with *Enterobacter hormaechei* (*bla*ACT-15 and *qnrB1*), *Enterobacteriaceae* bacterium (*blaTEM-1B* and *dfrA14*), *Salmonella enterica* (*aph(3″)-Ib*, *aph(6)-Id*, *dfrA14*, *sul2*), *Solibaculum mannosilyticum* (*tet(W)*), and *Streptococcus equinus* (*lnu(C)*) prior to HSCT, while non-GvHD patients had *Enterococcus faecalis* (*erm(B)*), *Staphylococcus aureus* (*lnu(B)* and *lsa(E)*), and *Streptococcus suis* (*ant(6)-la*). One week after HSCT, only *E. faecium* with various combinations of *msr(C)*, *aac(6′)-li*, *erm(B)*, *ant(6)-la*, *dfrG*, *VanHAX*, *erm(T)* and *aph(3′)-III* resistance genes were detected in both groups.

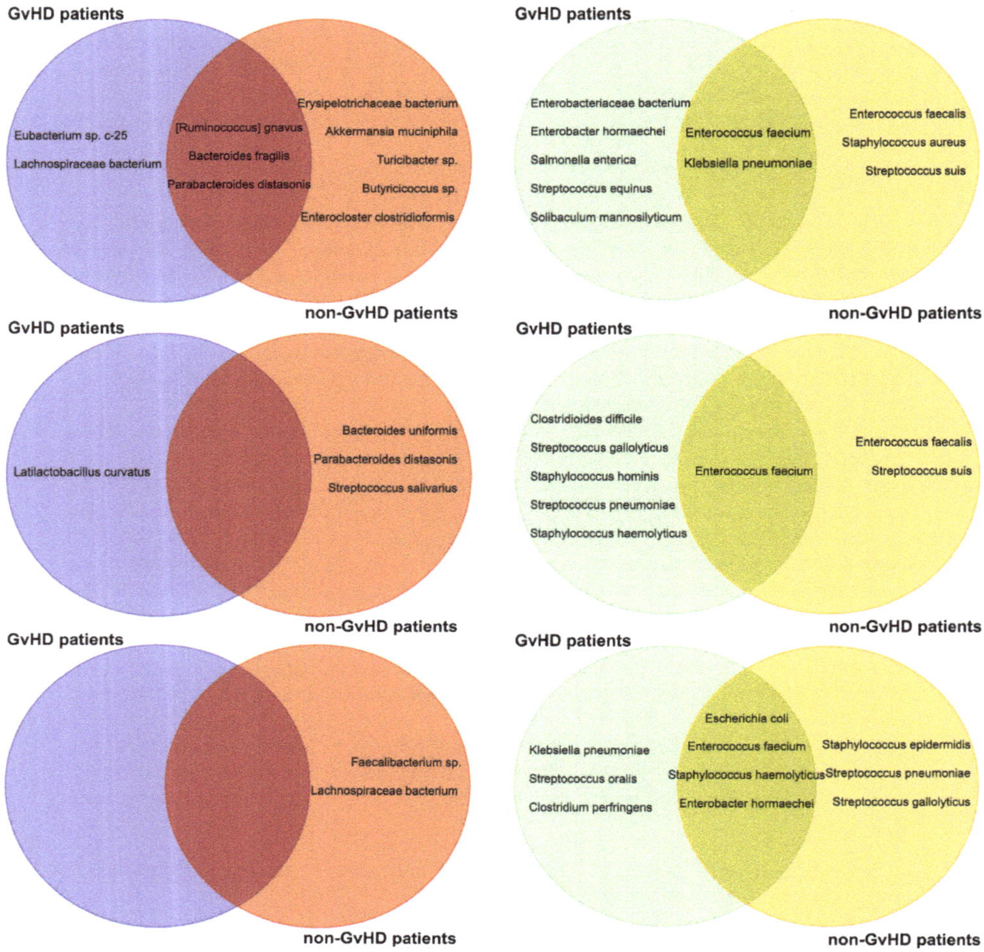

Figure 4. Schematic visualization of multidrug-resistant commensal and opportune pathogen bacteria identified in the gut microbiome of pediatric oncology patients suffering from acute lymphoblastic leukemia developing GvHD. d − 7—before HSCT, d + 7—1 week after HSCT, d + 28—1 month after HSCT.

The patients who did not develop GvHD remained resistant even one week after HSCT. However, those who did develop GvHD had an increased relative abundance of opportunistic bacteria such as *S. haemolyticus*, *Clostridioides difficile*, *Streptococcus gallolyticus*, *Staphylococcus hominis*, and *Streptococcus pneumoniae* present in their system. One month after HSCT, both GvHD and non-GvHD patients had *E. faecium*, *S. haemolyticus*, *E. hormaechei*, and *Escherichia coli* present in their system. *K. pneumoniae* and *S. suis* were specific to GvHD patients, while *Staphylococcus epidermidis*, *S. gallolyticus*, and *S. pneumoniae* were specific to non-GvHD patients one month after HSCT. We investigated the antimicrobial-resistance gene profiles of MDR bacteria, concluding that before HSCT most patients carried resistance genes *sul2*, *blaTEM-1B*, *aph(3)-Ib* by *Klebsiella pneumoniae*, *aac(6)-Ii*, *msr(C)*, *aac(6)-aph(2)*, and *ant(6)-Ia* by *Enterococcus faecium*, while one week after HSCT, as well as one month after HSCT, *Enterococcus faecium* carrying *aac(6)-Ii*, *ant(6)-Ia* and *drf(G)* dominated unequivocally (Figure 5).

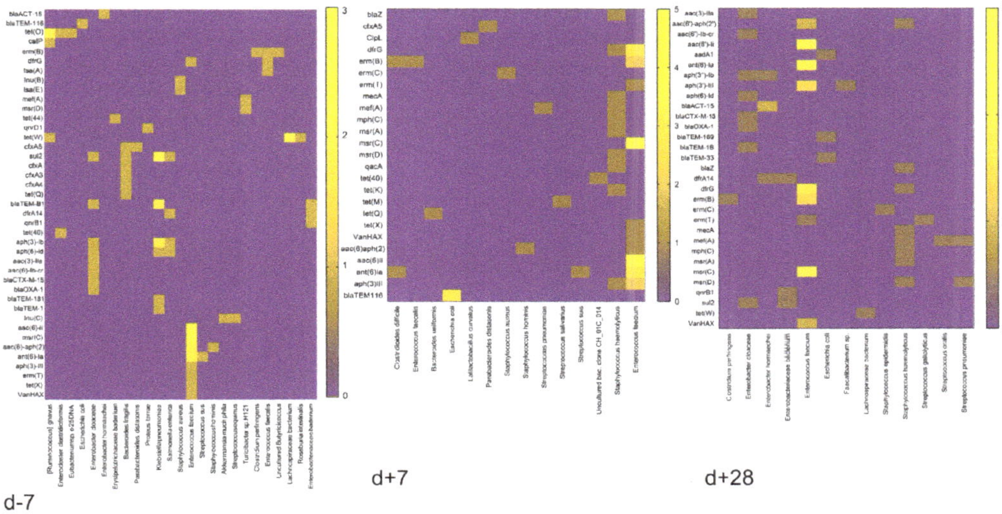

Figure 5. Schematic heatmap visualization of antimicrobial-resistance gene profiles identified in MDR bacteria of all patients. The color represents the number of patients with identified antimicrobial resistance genes carried by specific gut bacteria.

These results highlight the differences in microbiome composition and antibiotic resistance gene profiles (Figure 5). The unique combination of bacterial strains and resistance genes in each patient may affect their response to antibiotic treatment and the development of GvHD. However, the results showed more variability in gene representation one week after HSCT in GvHD patients. This could be a marker and a timepoint for the detection and determination of the risk associated with GvHD development.

2.7. The Gut Bacteriome of Pediatric Oncology Patients Characterized by the 16S rRNA Analysis

The most abundant bacterial phylum identified by 16S rRNA sequencing was *Bacillota*, with an abundance of 69% in patients with GvHD and 79% in patients without GvHD before HSCT. Patients with GvHD showed a trend of increased relative abundance of *Bacillota* one week (71%) and one month (86%) after HSCT. In contrast, patients without GvHD showed different patterns with decreased relative abundance (72%) one week after HSCT and a significant, faster, and more remarkable increase (99%) one month after HSCT.

Interestingly, for patients with GvHD, the relative abundance of the second most abundant phylum, *Bacteroidota*, decreased from 19% before HSCT to 3% one week and 0.5% one month after HSCT. However, patients not developing GvHD possessed a relatively stable relative abundance of *Bacteroidota* before HSCT (11%) and one week after HSCT (12%), while a significant decrease (0.4%) could be observed one month after HSCT.

Pseudomonadota was found to be more prominent in patients who developed GvHD (graft-versus-host disease). Prior to HSCT (hematopoietic stem cell transplantation), the abundance of *Pseudomonadota* was 12% in patients with GvHD and 3% in those without GvHD. One week after HSCT, patients with GvHD showed a significant increase in the abundance of *Pseudomonadota*, which reached 23%, whereas patients without GvHD maintained an abundance of 8%. However, one month after HSCT, the abundance of *Pseudomonadota* decreased to 11% in patients with GvHD and 1% in those without GvHD.

Regarding *Verrucomicrobiota*, which had an abundance of less than 1% in all three time points for patients developing GvHD, despite the same abundance before HSCT in the cohort of non-GvHD patients, one week after HSCT, its abundance increased to 8%, suggesting its protective activity of the gut mucus layer.

Actinomycetota was present in all timepoints for both groups, with an abundance of 0.5% before HSCT for patients with GvHD and 1% for patients without GvHD. However, one week after HSCT, there was an increase in the relative abundance of *Actinomycetota* in GvHD patients (4%), while the percentage for patients without GvHD decreased to 0.3%. One month after HSCT, the abundance of *Actinomycetota* in patients without GvHD reached only 0.07%, whereas it was elevated to 2% in the gut of patients with GvHD. *Campylobacterota* was present only before HSCT in both groups, with an abundance of less than 1%. The other bacterial phyla did not show any significant change in relative abundance. (Figure 6).

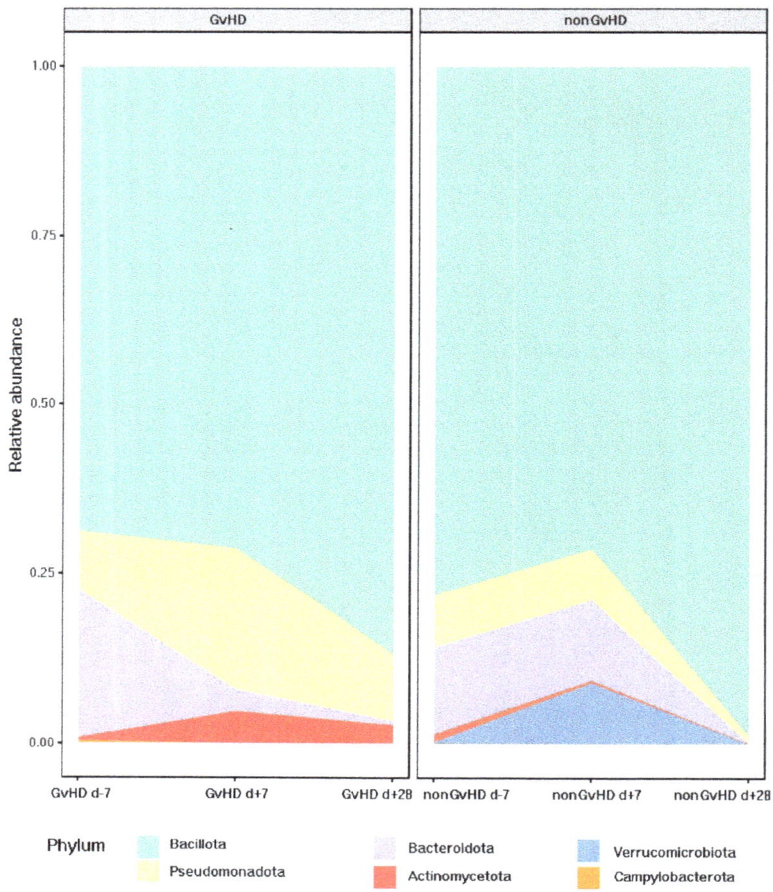

Figure 6. The gut microbiome of patients with acute lymphoblastic leukemia developing GVHD and the relative abundances of bacterial phyla.

Enterococcaceae was found to be the most abundant family across all timepoints in both groups. In the cohort of patients with GvHD, it had an abundance of 25% in all three timepoints, while in patients without GvHD, it showed an increase in abundance over time, i.e., 33% before HSCT, 52% one week after HSCT, and 79% one month after HSCT. *Bacteroidaceae* showed abundances of 18% in the gut of patients with GvHD and 10% in patients without GvHD before HSCT, with a decrease one week after HSCT (0.4%—GvHD+ patients/0.1% GvHD− patients). The abundances of the two groups differed one month after HSCT, where patients with GvHD had an average of only 0.4%, while patients without GvHD had an abundance of 0.03%. *Lachnospiraceae* was the third most abundant family before HSCT in both groups (13%—GvHD+ patients/10%—GvHD− patients), followed by a decrease to 7% one week after HSCT and <1% one month after HSCT in both groups. *Enterobacteriaceae* showed an increasing trend in patients with GvHD, from 10% before HSCT to 12% one week after HSCT, and a decrease to 8% one month after HSCT. On the other hand, patients without GvHD showed a decrease from 0.4% before and one week after HSCT to 0% one month after HSCT. A decreasing trend was observed in *Erysipelotrichaceae, Erysipeloclostridiaceae, Ruminococcaceae,* and *Veillonellaceae.*

The most abundant genus was Enterococcus across all timepoints in both groups, with stable abundance in all timepoints in patients with GvHD (25%) and with an increasing trend in patients without GvHD (33%/52%/79%). *Bacteroides* decreased in abundance over time for GvHD patients, with abundances of 18% before HSCT, 0.4% one week after HSCT, and 0.3% one month after HSCT. The gut microbiota of patients without GvHD showed a decrease from 11% before HSCT to 0.1% one week after HSCT and 0.03% one month after HSCT. *Erysipelatoclostridium,* together with *Lachnoclostridium, Faecalibacterium, Blautia,* and *Roseburia,* gradually decreased in both groups. *Staphylococcus* was higher in the samples one week and one month after HSCT in patients with GvHD, where the abundance increased from 0.3% before HSCT to 20% one week after HSCT and 15% one month after HSCT. The abundances in patients without GvHD were 16% before HSCT, 3% one week after HSCT, and 14% one month after HSCT. Similarly, this trend was observed in *Streptococcus,* increasing in patients with GvHD (0.05%/5%/12%) but decreasing in patients without GvHD (7%/0.3%/3%). *Klebsiella* was another bacterial genus enriched in the gut of patients with GvHD (0%/4%/8%) compared to patients without GvHD, where it was not detected. *Veillonella* was more abundant in the gut of patients with GvHD before HSCT, with an abundance of 3%, while at the other timepoints it was under 1%, similar to what we observed at all timepoints for the patients without GvHD.

The most abundant species identified in metagenomic data from both groups was *Enterococcus faecium* at all timepoints. The individual patients showed a wide range of abundances; therefore, a heatmap was used to better visualize the portfolio and the spectrum of individual bacterial species (Figure 7).

The following dominant pathobionts were detected: *E. faecium, E. faecalis, E. bolteae, S. haemolyticus, P. aeruginosa,* and *K. pneumoniae.* In addition, also commensal bacteria were detected in high abundance, including *Akkermansia muciniphila, Bacteroides uniformis, Parabacteroides distasonis, Bacteroides fragilis,* and *Latilactobacillus sakei* (Figure 7).

We found that multiresistant bacteria were dominant or had a higher abundance within the group of resistant bacterial species (Table 1). For instance, the relative abundance of a multiresistant strain *Ruminococcus gnavus* was 26%, as identified by 16S rRNA analysis and confirmed by whole metagenome sequencing (23%), making it the most abundant bacteria in patients' gut microbiota. For patients with dominant multiresistant *S. haemolyticus,* these bacteria reached 60% and 73% of total relative abundance, respectively. An analysis of the gut microbiome revealed the presence of multiresistant bacteria that were not identified by cultivation analysis, except for two patients who were found to be positive both in cultivation analysis of sputum or tonsils swabs as well as sequencing analysis (*E. faecium* KUHS13 and ME3 strains) of the gut. Patients with *Enterobacter* sp. multidrug resistance strains showed only a higher abundance of the genus *Enterobacter,* while the species from this genus were not well identified. Our analysis of the metagenomes revealed the consequences of the multidrug-resistant bacteria and their colonization of the disrupted gut microbiota environment (Figure 8).

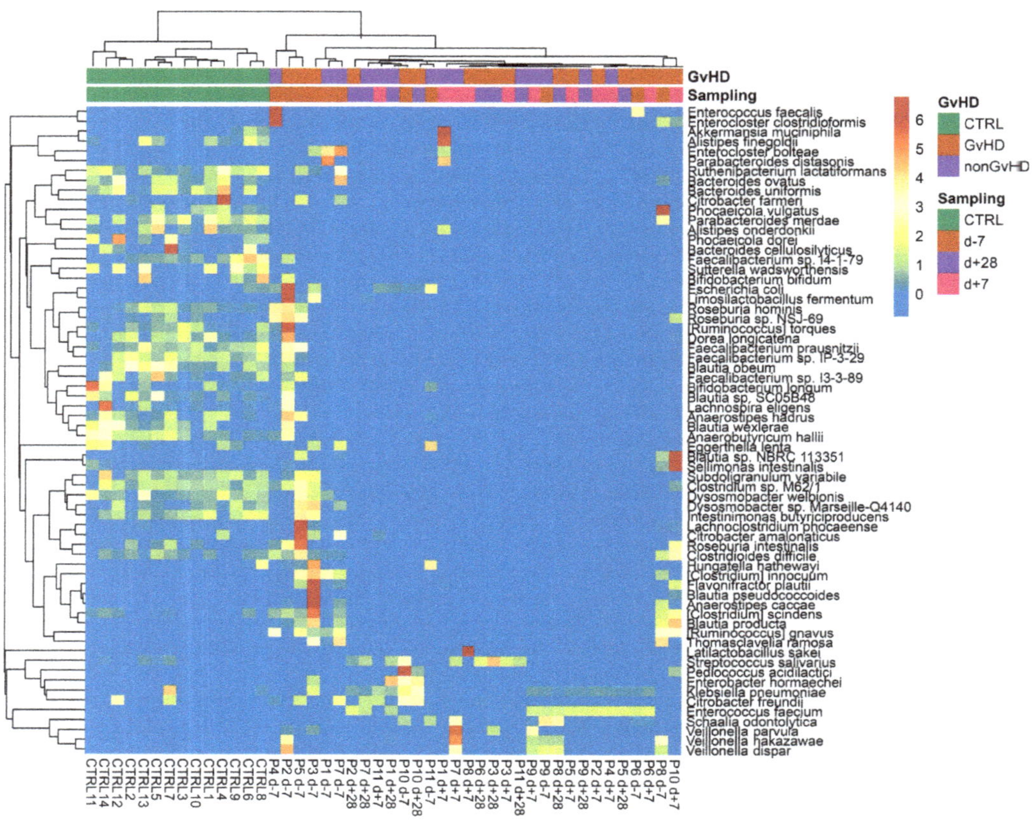

Figure 7. Graphical visualization of the 70 most abundant bacterial species (relative abundance) of the gut microbiome of pediatric oncology patients suffering from acute lymphoblastic leukemia using heatmap.

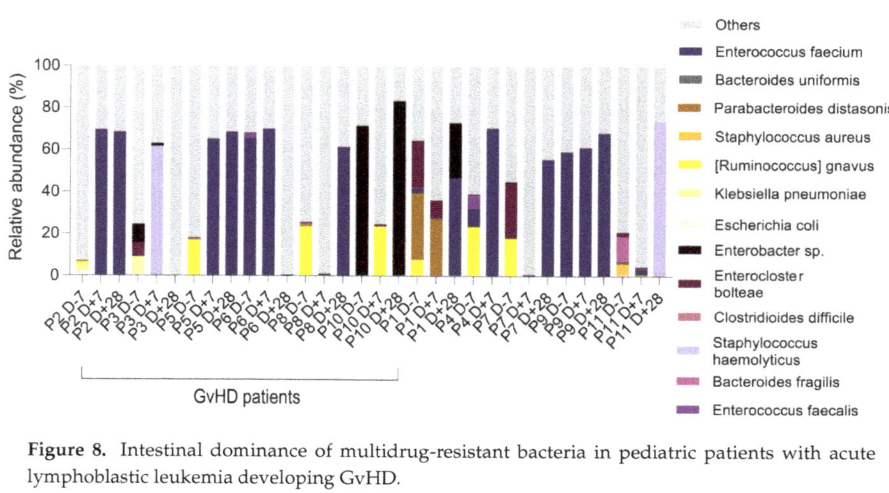

Figure 8. Intestinal dominance of multidrug-resistant bacteria in pediatric patients with acute lymphoblastic leukemia developing GvHD.

2.8. Development of Gut Mycobiome of Pediatric Patients Suffering from Acute Lymphoblastic Leukemia during HSCT Treatment

The median abundance of *Fungi* found in the samples was 0.03%. The highest abundance was 4.6%, while the lowest was 0.001% (with a standard deviation of 0.73%). The relative abundance was recalculated, with the final abundance of *Fungi* being 100%.

The mycobiome was characterized from metagenomic data; the most abundant fungi found were *Saccharomyces cerevisiae,* with average abundances of 44%/50%/50% in GvHD patients and 28%/18%/65% in non-GvHD patients (Figure 9). Healthy controls had an abundance of 7%. The second most abundant fungi in the gut microbiota of patients with GvHD before HSCT was *Yarrowia lipolytica,* with an abundance of 3%, which decreased to less than 1% one week and one month after HSCT. In patients without GvHD, the abundance was less than 1% at all three time points, while healthy controls had an abundance of 1%. *Nakaseomyces glabratus* (*Candida glabrata*) and *Malassezia restricta* (Figure 9) were identified with abundances of 0.1% and 3%, respectively, in GvHD patients. The abundance of *N. glabratus* increased to less than 0.1% one week after HSCT and one month after HSCT. In non-GvHD patients, the abundance of *N. glabratus* increased from less than 1% before HSCT to 2% a week after HSCT and 13% a month after HSCT (Figure 9). The abundance in healthy individuals was less than 1%. *M. restricta* in patients with GvHD increased to 2% one week after HSCT and 5% one month after HSCT. The abundances in patients without GvHD were 3%/8%/1%, while healthy controls had less than 1%. Other abundant species found in patients were *Fusarium keratoplasticum* (0.3%/0.1%/0.01% in GvHD patients and 0.1%/0.1%/0.1% and 1% in healthy controls), *Candida orthopsilosis* (0.1%/0.1%/6% in GvHD patients; 1%/0.1%/0.0003% in non-GvHD patients; 1% in healthy controls), *Candida dubliniensis* (0.1%/0.04%/0.2% in GvHD patients; 2%/2%/0.5% in non-GvHD patients; 1% in healthy controls), and *Neurospora crassa* (0.3%/0.2%/0.6% in GvHD patients; 1%/0.1%/0.6% in non-GvHD patients; 2% in healthy controls).

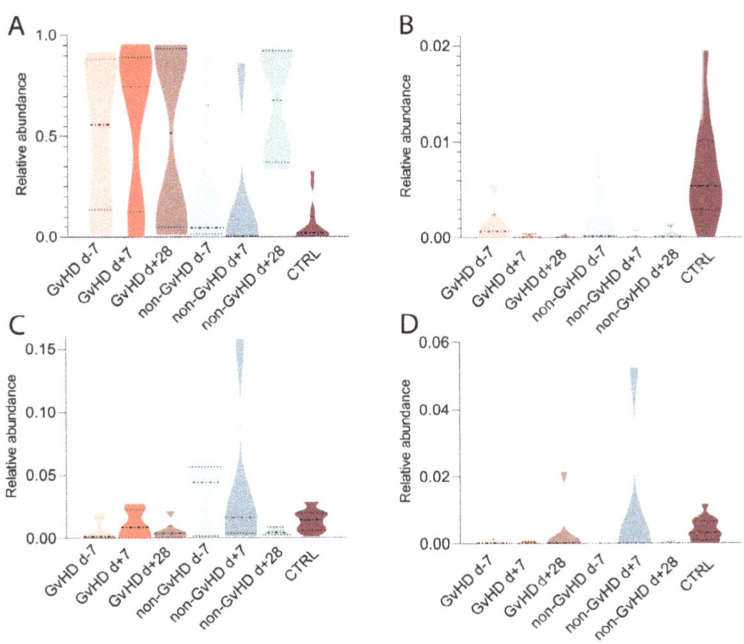

Figure 9. Violin plots showing the relative abundances (%) (recalculated to the Fungi—100%) of selected fungal species identified within the gut of patients developing GvHD. The median with the first and third quartile are marked. (**A**)—*Saccharomyces cerevisiae;* (**B**)—*Malassezia restricta;* (**C**)—*Nakaseomyces glabratus;* (**D**)—*Yarrowia lipolytica.*

Overall, there were no significant differences in the relative abundance of the most abundant fungal species. A statistical analysis allowed us to identify *Marasmius* ($p = 0.004$) as the only differentially abundant fungus that was significantly more abundant in the gut of GvHD patients.

2.9. The Alpha Diversity of the Gut Microbiome of Patients with GvHD

The α-diversity was evaluated using the Shannon and Simpson indices. In the cohort of patients developing GvHD, a decreasing trend in diversity richness and evenness was observed. In contrast, patients without GvHD showed a repeated diversity increase one month after HSCT. The Shannon index (average ± stdev) was determined to be $2.7 \pm 1/1.6 \pm 0.6/1.3 \pm 0.8$ in GvHD patients and $2.1 \pm 0.7/1 \pm 0.5/1.7 \pm 1$ in non-GvHD patients. The Simpson index was determined to be $0.8 \pm 0.2/0.6 \pm 0.1/0.5 \pm 0.2$ in GvHD patients and $0.7 \pm 0.2/0.4 \pm 0.1/0.6 \pm 0.3$ in non-GvHD patients (Figure 10). The alpha diversity of the gut microbiome of healthy children was significantly ($p < 0.01$) higher (Shannon index 4.1 ± 0.4; Simpson index 0.9 ± 0.1) than that of oncology patients, both either developing or not GvHD. However, no significant changes in diversity between patients developing GvHD and non-GvHD patients were observed, either before or after HSCT treatment.

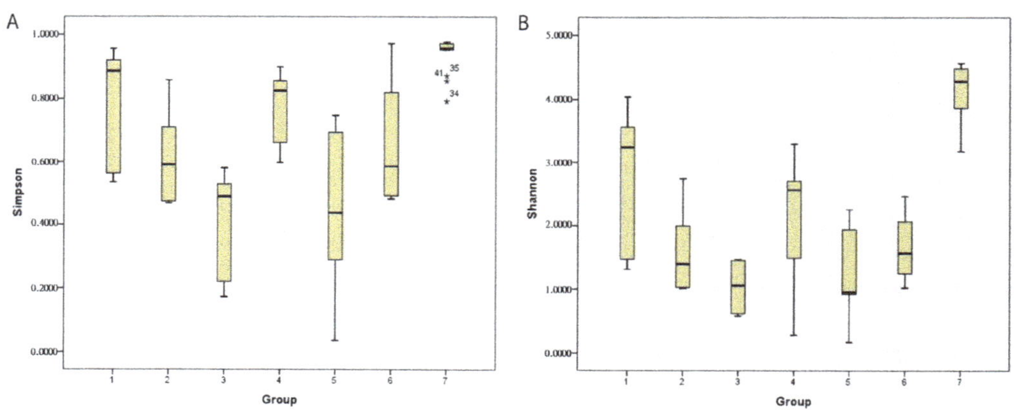

Figure 10. The alpha diversity of the gut microbiome of pediatric patients who have acute lymphoblastic leukemia is represented by Simpson's (**A**) and Shannon's (**B**) indices. 1—GvHD patients d − 7; 2—GvHD patients d + 7; 3—GvHD patients d + 28; 4—non-GvHD patients d − 7; 5—non-GvHD patients d + 7; 6—non-GvHD patients d + 28; 7—healthy controls; *—outlier.

Nevertheless, we found that the richness of the gut microbiome in GvHD patients one month after HSCT was significantly lower compared to that before HSCT ($p = 0.037$) and one week after HSCT ($p = 0.002$).

3. Discussion

We conducted a metagenomic analysis to investigate the gut resistome of pediatric patients undergoing HSCT and healthy controls and to observe changes in the gut microbiome composition and diversity. As far as we know, the only study that analyzed the gut resistome of GvHD patients was conducted by D'Amico et al. (2019), who found a broad pattern of resistance classes and post-HSCT expansion of the gut resistome in eight pediatric patients [12]. Our study specifically focused on patients with ALL, of whom six had TBI, and five had a chemotherapy-based conditioning regimen. We found that all the factors and treatments that patients routinely undergo can significantly impact microbial composition and diversity. *Klebsiella*, *Enterococcus*, *Staphylococcus*, and *Enterobacter* are part of the human gut microbiota and represent infectious agents. Patients are more susceptible to these bacteria [13]. Several studies have described *Enterococcus* as the main biomarker of

GvHD development [8,14,15]. However, we could not confirm this in our study. On the contrary, we found that *Enterococcus* abundance was higher in non-GvHD patients. This could be due to the limited number of patients in our study, whereas other studies had significantly larger cohorts. Our analysis suggests that *Enterococcus* may pose a greater risk to younger children, as its abundance was significantly higher in children under ten years of age. Devaux et al. (2020) described a possible link between GvHD, lactose accumulation in the gut lumen, Reg3 reduction, and lactose auxotroph *E. faecium* proliferation. Due to antimicrobial treatment, depletion in gut microbiota reduced Actinobacteria, which cross-fed the butyrate-producing bacteria from the *Clostridia* class and decreased the metabolism of acetate to butyrate [13]. Butyrate has been described as a protector in GvHD development. All these factors can lead to *Enterococcus* dominance and inflammatory processes, with higher expression of myosin light chain kinase 210 (MLCK210) and subsequent disruption of the intestinal barrier and translocation of bacteria from the gut. Even if it was not found to be multi-drug resistant, *Enterococcus* was found to be the dominant bacteria in most patients.

The gut resistome plays an essential role in medical research, especially for immunocompromised patients [14]. A comparison of the resistance genes present in HSCT patients' gut microbiome versus healthy controls showed that HSCT patients exhibited a more frequent and broader spectrum of resistances. Moreover, pre-HSCT samples demonstrated a broad spectrum of resistance genes, likely due to exposure to the hospital environment and previous high-dose chemotherapy treatments [15]. Pediatric patients demonstrated an especially pronounced expansion of the gut resistome one week after HSCT in GvHD patients. Since patients undergoing HSCT require immunosuppressive therapy, antimicrobial treatment is often necessary to reduce the risk of infection. However, antibiotics can exert selective pressure, favoring the expansion of antibiotic-resistant strains within the gut microbiota. Despite the use of broad-spectrum antibiotics before HSCT (d − 10), multiresistant bacteria, often with multidrug resistance genes, can survive the combined antimicrobial treatment and lead to therapeutic challenges. Hence, the presence of multidrug resistance genes in the gut microbiota can make common antibiotics less effective.

Our study has highlighted the presence of multiresistant bacteria in the gut microbiota of patients, while the healthy controls showed minimal presence of resistance genes in their gut microbiome. Patients' gut microbiota also exhibited resistance genes before HSCT due to a strict antibiotic treatment [16]. Resistance genes, which are suggested to be beneficial for our bodies, were detected in commensal bacteria of non-GvHD patients. *Bacteroidales* representatives were most common, but *Lachnospiraceae*, *Clostridiaceae*, *Erysipeltorichales*, and *Akkermansia* were also detected. *Akkermansia* is a bacterial genus often correlated with maintaining the gut barrier and protecting the gut [17]. The gut barrier is especially important in GvHD, where its disruption can lead to the translocation of bacteria into the bloodstream, increasing the risk of infection development. Evidence suggests that conditioning regimen-induced gut barrier damage is the primary driver for GvHD development [18–20]. Although commensal bacteria play an essential role in maintaining gut homeostasis and balance, carrying resistance genes could pose a risk for the horizontal transfer of bacterial genes to pathogenic bacteria within the gut or opportunistic pathogens from external sources [21]. Therefore, while the presence of resistance genes in commensal bacteria may not pose an immediate and severe threat, it does highlight the need for careful monitoring and a deeper analysis of the dynamics of the gut microbiome.

MDR bacteria have developed resistance to multiple antibiotics, which makes them more potent in the unfavorable environment created by antimicrobial prophylaxis and conditioning regimens. The colonization of the human body by MDR bacteria could be the most common source of infectious microorganisms; this has been observed to occur more frequently in GvHD patients [22]. Another study found that infection by MDR gram-negative bacteria significantly reduced overall survival one year after HSCT. While MDR bacteria pose a potential danger to patients, deaths associated with MDR bacteria are rare. However, they still contribute to complicated and prolonged hospitalizations. The mechanism by which opportunistic pathogens can affect GvHD development is still unclear. Still, some studies have suggested that these pathogens in the human gut microbiome are prone to acquiring resistance and surviving rigorous antimicrobial therapy. There are supporting hypotheses that may explain how opportunistic pathogens can support GvHD development. One possibility is gut barrier disruption, as some bacterial pathogens can impair gut barrier functions, weaken tight junctions by disrupting zonula occludens, occludin, and claudin, and initiate inflammation [23]. Inflammation is a second possible mechanism, as it can stimulate host cells through microbe-associated molecular patterns, which could enhance inflammation in GvHD [24,25].

Our investigation has revealed important findings that suggest the need for more effective strategies to prevent infection and colonization by multiresistant bacteria. We observed that the combination of broad-spectrum targeted antimicrobial treatment with broad-spectrum antibiotics may not be sufficient to avoid the presence of multiresistant bacteria, which were dominant or significantly more abundant in individual patients. Furthermore, this combination of antibiotics was associated with a loss in diversity and an enrichment in the presence of gut bacteria with resistance genes. These findings highlight the need for more effective and targeted approaches to managing infections in immunocompromised patients, as well as the importance of monitoring the dynamics of the gut microbiome. We have identified shifts in the gut microbiota composition between GvHD and non-GvHD patients, consistent with previous studies. We identified elevated levels of facultative anaerobes from the *Bacillota* phylum, such as *Enterococcus* and *Staphylococcus*, along with a decrease in *Bacteroidota* [26]. However, the time frame for these changes differed between the two groups. GvHD patients experienced a rapid reduction in *Bacteroidota* abundance just one week after HSCT, while non-GvHD patients experienced this shift one month after transplantation. Additionally, the *Pseudomonadota* phylum showed varying abundance between the two groups, with GvHD patients displaying higher levels. The representants of this phylum were detected in higher abundance in the HSCT patients [27]. Despite these differences, we also noticed a commonality in the loss of SCFA-producing bacteria across all patient groups compared to healthy controls, predominantly from the *Clostridia* class. These findings align with previous research, indicating the consistency of these alterations in the context of HSCT [28–30]. These observations have important implications for future research and clinical practice, as they suggest that restoring SCFA-producing bacteria in the gut may be a key strategy in promoting gut health and reducing the risk of complications after HSCT.

Furthermore, our study has identified a substantial decrease in alpha diversity among transplant recipients compared to healthy controls, reflecting the reduced overall diversity in the gut microbiota of these patients. We observed that GvHD patients experienced a more pronounced loss of diversity over time, which may contribute to the pathogenesis of GvHD and other health issues. These findings highlight the need to develop interventions that promote diversity and improve outcomes in transplant recipients. Further research is necessary to identify the factors contributing to decreased diversity in the gut microbiota of these patients and to develop targeted interventions based on these factors.

The role of the gut mycobiome in acute GvHD has been described as showing involvement of *Candida* spp., primarily represented by *Candida albicans*, in the development of aGvHD with higher severity and more often developed gastrointestinal aGvHD [31]. Not the diversity of gut mycobiome, but *Candida* spp. abundance at day 0 of HSCT has been shown to play a crucial role in the overall survival of patients, as well as GvHD development after alloHSCT, probably triggered by β-d-glucan and mannans, leading to the induction of Th 17 and IL-23 response in the mucus layer [32]. Currently, however, only studies focused on adult oncology patients have been published; notably, most of the patients in these studies did not receive antifungal therapy [31,32]. Our results, therefore, provide unique information on the gut mycobiome of pediatric oncology patients suffering from acute lymphoblastic leukemia undergoing HSCT with a focus on GvHD development. Although there was a limited number of patients in our study, *Candida* spp. was represented exclusively by *Nakaseomyces glabratus* (*Candida glabrata*) and *Candida dubliniensis*, but not *Candida albicans*. Both fungal species were more abundant in the gut of patients not developing GvHD either before HSCT or one week after HSCT, suggesting that this represents a biomarker of better HSCT outcome and higher microbiome diversity.

The main limitation of our study was the low number of patients. However, we included all pediatric oncology patients admitted to the hospital during three consequent years, provided they were able to provide a biological specimen for gut microbiome analysis at all three established timepoints. Additionally, so far, there is only one scientific paper focused on resistome profiling of patients with GvHD, so our findings represent unique results that cannot be compared to previously published articles.

We are aware of the heterogeneity of patients, and that is why we performed analyses on confounding factors that could possibly contribute to patient gut microbiome variability. Even though the subgroups were smaller, we aimed to clearly identify the potential effect of confounding factors and pinpoint their potential role in GvHD outcome. If no significant differences in gut microbiome composition or diversity were observed, we estimated the variability of sample groups without any significance to the measured parameter.

Regarding the limitations of the bioinformatics tools and databases, a plethora of different approaches are used for microbiome analysis; therefore, we used the most relevant bioinformatic tools and approaches (to the best of our knowledge) for microbiome 16S rRNA and metagenomic analysis, including the Silva database used with QIIME2 and DADA2, as well as Kraken2, with the standard database, including relevant and reliable representative genomes of microorganisms for taxonomical classification. However, we admit that analyses of the other variable regions of 16S rRNA may lead to similar but alternative results. Another possible limitation involved the ResFinder database, which does not cover the resistance mechanisms of chromosomal mutations and only consists of known and described resistance gene sequences. On the other hand, the vfdb database for virulence genes is limited to only 32 bacterial genera classified as pathogens.

We nonetheless believe that our study provides valuable insights into the dynamics of the gut microbiota in transplant recipients and can guide future research and clinical practice in this area.

4. Materials and Methods

4.1. Subjects

In total, 32 patients were involved in the present study for gut microbiome analysis. Each patient provided one to three biological specimen samples. For paired analyses, 61 fecal samples were obtained from 21 randomly assigned pediatric oncology patients at the Department of Paediatric Haematology and Oncology-Bone Marrow Transplant Unit (Faculty of Medicine at Comenius University and National Institute of Child Diseases) from 2019 to 2021. The samples were collected seven days before HSCT (d − 7), one week (d + 7), and one month (d + 28) after HSCT.

The cohort of patients included 11 males and seven females from three to 18 years of age (11.8 ± 5.2 years). Eleven patients were diagnosed with acute lymphoblastic leukemia, three with acute myeloid leukemia, two with myelodysplastic syndrome, one with anaplastic large cell lymphoma, and one with X-linked lymphoproliferative disease. Twelve patients developed febrile neutropenia (FN), and nine patients were diagnosed with GvHD; six had been diagnosed with both GvHD and FN. Of the eleven patients (five females and six males) with acute lymphoblastic leukemia, six were treated in a conditioning regimen with total body irradiation (TBI) and five with chemotherapy; five patients developed GvHD. Of the five patients undergoing chemotherapy, three developed GvHD, while three of the six patients with TBI developed GvHD. All patients developed GvHD from day 16 after HSCT to day 35. The legal representatives of all participants signed informed consent. Samples from fourteen healthy children (six males, eight females) aged 4 to 18 (8.5 ± 4.4 years) were collected in 2021. Randomly assigned healthy individuals had not been diagnosed with any acute or chronic disease and had not received any antimicrobial treatment for one month before sampling. The legal representatives of all participants signed informed consent forms. For the purposes of the present study, patients, that provided at least two biological specimens, obtained within the period of HSCT therapy, were included.

4.2. V1-V3 16S rRNA Gene Amplicon Library Preparation and Bioinformatic Analysis

DNA was isolated using a Zymobiomics DNA/RNA isolation kit. The nucleic acids were isolated according to the manufacturer's protocol. DNA was used for V1-V3 16S rRNA amplicon generation and whole-genome shotgun sequencing. The amplicon PCR was prepared using 5× HotFirePOL Mastermix (Solis BioDyne, Tartu, Estonia) with a primer pair from the Human Microbiome Project (https://www.hmpdacc.org/, accessed on 1 November 2023) and a 30-cycle PCR program. The PCR products were then indexed with the low-cycle protocol using the Nextera XT Index kit (Illumina, San Diego, CA, USA).

The concentration of the final libraries was measured by Qubit 4.0 Fluorometer (Thermo Fisher Scientific, Waltham, MA, USA), and the DNA profile of the obtained sequences was determined with an Agilent Bioanalyzer using DNA High Sensitivity Kit (Agilent Technologies, Santa Clara, CA, USA). The final libraries were sequenced using the Illumina MiSeq platform via300 bp paired-end reads (Illumina, San Diego, CA, USA).

Samples were analyzed using the Qiime2 2022.11 pipeline [33]. Imported paired-end sequences with quality data were corrected and filtered using the DADA2 QIIME2 plugin with the following parameters: --p-trunc-len-f 285, --p-trunc-len-r 285, --p-trunc-q 2, and --p-max-ee-r 3 [34]. Output from DADA2 was used for further taxonomic classification. For the taxonomic classification of the V1-V3 region of 16S rRNA, we used the pre-trained Silva 138 classifier in the q2-feature-classifier function [35,36]. OTUs classified with the Qiime2 2022.11 pipeline were used to create a sample-specific bacterial profile.

4.3. Whole-Genome Shotgun Sequencing and Bioinformatic Analysis

DNA libraries were prepared using a Nextera kit based on transposon tagmentation and PCR. For metagenomic analysis, total DNA was fragmented using transposon-based chemistry (Illumina, San Diego, CA, USA); the obtained fragments were further indexed using low cycle PCR with a Nextera XT Index kit according to manufacturer's instructions (Illumina, San Diego, CA, USA). Final DNA libraries were checked with a Qubit 4.0 Fluorometer (Thermo Fisher Scientific, Waltham, MA, USA) for quantity and chip electrophoresis using an Agilent Bioanalyzer 2100 system (Agilent Technologies, Santa Clara, CA, USA) with a High Sensitivity DNA Kit to assess quality. Samples were sequenced on the Illumina NextSeq (Illumina, San Diego, CA, USA) platform (2 × 150 bp). All steps in the methodology were performed according to the manufacturer's protocol.

The reads underwent pre-processing, involving quality control and trimming, through an in-house Galaxy web instance [37]. Subsequently, quality reports for the reads were generated using the FastQC tool [38]. Trimming was executed using Trimmomatic—Galaxy Version 0.38.1 with the following parameters: CROP:296, HEADCROP:17, MINLEN:50 [39].

For metagenomic assembly, the metaSPAdes v3.15.4 software was employed, utilizing the default settings [40]. To classify the taxonomical information of the trimmed reads and metaSPAdes scaffolds, Kraken 2 was used in conjunction with a custom-built database. This database incorporated information on archaea, bacteria, plasmids, viruses, humans, fungi, protozoa, and UniVec_core Kraken 2 sequence libraries, which were retrieved on 8 March 2023. The creation of this database followed the Kraken 2 protocol [41].

4.4. Identification of the Drug Resistance and Virulence Genes

Sequences assembled to contigs were used as the input for the ABRIcate tool Galaxy version 1.0.1 (https://github.com/tseemann/abricate) with the default settings. The ResFinder database was used for the detection of resistance genes, and the VFDB database was used for virulence gene identification [42]. The minimum DNA identity was 97%, and the minimum DNA coverage was 100%. The presence/absence of a gene expressed as 1/0 was used for further statistical analysis. The individual contigs with detected resistance genes were then analyzed through BLAST (megablast), which allowed the identification of bacterial strains carrying the antimicrobial-resistance genes [43]. The bacterial strains with the highest identity and coverage score were considered.

4.5. Statistical Analysis

Data normality was determined by SPSS (IBM Corp. Released 2012. IBM SPSS Statistics for Windows, Version 21.0. Armonk, NY, USA: IBM Corp.) using the p-value of the Shapiro-Wilk test. For further statistical analysis of the differential abundance of bacterial and fungal taxa, the LefSe tool based on the Kruskall-Wallis test was applied [44]. ClustVis tool BETA version (https://biit.cs.ut.ee/clustvis/) was used for heatmap visualization, and Gephi version 0.9.7 (https://gephi.org/) was used for the network visualization of the individual resistance genes for the specific antimicrobials. GraphPad (GraphPad Prism version 8.0.0 for Windows, GraphPad Software, San Diego, CA, USA, www.graphpad.com) was used for other graphical visualizations (bar charts, heatmaps), and the stack-area plot was made using the Microbiome analyst online tool [45].

Author Contributions: Conceptualization, K.S., P.S. and A.K.; methodology, K.S., G.B., L.K. and G.M.; software, G.B. and L.K.; validation, K.S., S.S. and K.A.; formal analysis, K.S., S.S., K.A. and G.M.; investigation, K.S., S.S. and K.A.; resources, K.S.; data curation G.B., K.S. and S.S., visualization, K.S., S.S. and K.A.; writing—original draft preparation, K.S., S.S. and K.A.; writing—review and editing, all authors; project administration, K.S. and V.B. All authors have read and agreed to the published version of the manuscript.

Funding: This research was funded by Slovak Research and Development Agency, grant. No. APVV-17-0099.

Institutional Review Board Statement: The study was conducted in accordance with the Declaration of Helsinki and approved by the Ethics Committee of the National Institute of Child Diseases (NICD-25/4/18).

Informed Consent Statement: Informed consent was obtained from all legal representatives of subjects involved in the study.

Data Availability Statement: The data represented in this study are openly available in the SRA database under BioProject ID PRJNA1009258.

Acknowledgments: We would like to acknowledge and are very grateful to all patients and their parents who have participated in this study and all doctors and nurses, who were responsible for sample collection, biobanking and categorization and of course professional treatment.

Conflicts of Interest: The authors declare no conflict of interest.

eferences

Gratwohl, A.; Brand, R.; Apperley, J.; Biezen, A.V.; Bandini, G.; Devergie, A.; Schattenberg, A.; Frassoni, F.; Guglielmi, C.; Iacobelli, S.; et al. Graft-versus-Host Disease and Outcome in HLA-Identical Sibling Transplantations for Chronic Myeloid Leukemia. *Blood* **2002**, *100*, 3877–3886. [CrossRef]

Jones, J.M.; Wilson, R.; Bealmear, P.M. Mortality and Gross Pathology of Secondary Disease in Germfree Mouse Radiation Chimeras. *Radiat. Res.* **1971**, *45*, 577–588. [CrossRef] [PubMed]

van Bekkum, D.W.; Roodenburg, J.; Heidt, P.J.; van der Waaij, D. Mitigation of Secondary Disease of Allogeneic Mouse Radiation Chimeras by Modification of the Intestinal Microflora. *J. Natl. Cancer Inst.* **1974**, *52*, 401–404. [CrossRef] [PubMed]

Bealmear, P.M.; Mirand, E.A.; Holtermann, O.A. Modification of Graft-vs-Host Disease Following Bone Marrow Transplantation in Germfree Mice. *Prog. Clin. Biol. Res.* **1983**, *132C*, 409–421. [PubMed]

Balletto, E.; Mikulska, M. Bacterial Infections in Hematopoietic Stem Cell Transplant Recipients. *Mediterr. J. Hematol. Infect. Dis.* **2015**, *7*, e2015045. [CrossRef]

Santos, E.S.; Lima, A.C.M.; Breda, G.L.; de Oliveira Tomaz, A.P.; Nabhan, S.K.; Funke, V.A.M.; Loth, G.; da Silva Nogueira, K. Colonization by Multidrug-Resistant Bacteria in Hematological Patients Undergoing Hematopoietic Stem Cell Transplantation and Clinical Outcomes: A Single-Center Retrospective Cohort Study. *Transpl. Infect. Dis.* **2023**, *25*, e14119. [CrossRef]

Heidenreich, D.; Kreil, S.; Nolte, F.; Hofmann, W.K.; Miethke, T.; Klein, S.A. Multidrug-Resistant Organisms in Allogeneic Hematopoietic Cell Transplantation. *Eur. J. Haematol.* **2017**, *98*, 485–492. [CrossRef]

8. Stein-Thoeringer, C.; Peled, J.U.; Lazrak, A.; Slingerland, A.E.; Shono, Y.; Staffas, A.; Gomes, A.; Argyropoulos, K.; Docampo, M.; Xavier, J.; et al. Domination of the Gut Microbiota with Enterococcus Species Early after Allogeneic Bone Marrow Transplantation Is an Important Contributor to the Development of Acute Graft-Versus-Host Disease (GHVD) in Mouse and Man. *Biol. Blood Marrow Transplant.* **2018**, *24*, S40–S41. [CrossRef]
9. Biagi, E.; Zama, D.; Nastasi, C.; Consolandi, C.; Fiori, J.; Rampelli, S.; Turroni, S.; Centanni, M.; Severgnini, M.; Peano, C.; et al. Gut Microbiota Trajectory in Pediatric Patients Undergoing Hematopoietic SCT. *Bone Marrow Transplant.* **2015**, *50*, 992–998. [CrossRef]
10. Doki, N.; Suyama, M.; Sasajima, S.; Ota, J.; Igarashi, A.; Mimura, I.; Morita, H.; Fujioka, Y.; Sugiyama, D.; Nishikawa, H.; et al. Clinical Impact of Pre-Transplant Gut Microbial Diversity on Outcomes of Allogeneic Hematopoietic Stem Cell Transplantation. *Ann. Hematol.* **2017**, *96*, 1517–1523. [CrossRef]
11. Ilett, E.E.; Jørgensen, M.; Noguera-Julian, M.; Nørgaard, J.C.; Daugaard, G.; Helleberg, M.; Paredes, R.; Murray, D.D.; Lundgren, J.; MacPherson, C.; et al. Associations of the Gut Microbiome and Clinical Factors with Acute GVHD in Allogeneic HSCT Recipients. *Blood Adv.* **2020**, *4*, 5797–5809. [CrossRef] [PubMed]
12. D'Amico, F.; Soverini, M.; Zama, D.; Consolandi, C.; Severgnini, M.; Prete, A.; Pession, A.; Barone, M.; Turroni, S.; Biagi, E.; et al. Gut Resistome Plasticity in Pediatric Patients Undergoing Hematopoietic Stem Cell Transplantation. *Sci. Rep.* **2019**, *9*, 564 [CrossRef] [PubMed]
13. Devaux, C.A.; Million, M.; Raoult, D. The Butyrogenic and Lactic Bacteria of the Gut Microbiota Determine the Outcome of Allogenic Hematopoietic Cell Transplant. *Front. Microbiol.* **2020**, *11*, 1642. [CrossRef] [PubMed]
14. Gill, J.S.; Singh, S.P.; Sharma, S.; Agarwal, A. Gut Microbiota Profiling in Hematopoietic Stem Cell Transplant Recipients: Towards Personalized Medicine. *Med. J. Armed. Forces India* **2021**, *77*, 22–27. [CrossRef] [PubMed]
15. High-Dose Chemotherapy and Stem Cell Transplant for Neuroblastoma. Available online: https://www.cancer.org/cancer/types/neuroblastoma/treating/high-dose-chemo-radiation.html (accessed on 31 October 2023).
16. Tanaka, J.S.; Young, R.R.; Heston, S.M.; Jenkins, K.; Spees, L.P.; Sung, A.D.; Corbet, K.; Thompson, J.C.; Bohannon, L.; Martin, P.L.; et al. Anaerobic Antibiotics and the Risk of Graft-versus-Host Disease after Allogeneic Hematopoietic Stem Cell Transplantation. *Biol. Blood Marrow Transplant.* **2020**, *26*, 2053–2060. [CrossRef] [PubMed]
17. Derrien, M.; Vaughan, E.E.; Plugge, C.M.; de Vos, W.M. Akkermansia Muciniphila Gen. Nov., Sp. Nov., a Human Intestinal Mucin-Degrading Bacterium. *Int. J. Syst. Evol. Microbiol.* **2004**, *54*, 1469–1476. [CrossRef]
18. Wang, Y.M.; Abdullah, S.; Luebbering, N.; Langenberg, L.; Duell, A.; Lake, K.; Lane, A.; Hils, B.; Vazquez Silva, O.; Trapp, M.; et al. Intestinal Permeability in Patients Undergoing Stem Cell Transplantation Correlates with Systemic Acute Phase Responses and Dysbiosis. *Blood Adv.* **2023**, *7*, 5137–5151. [CrossRef]
19. Clayburgh, D.R.; Barrett, T.A.; Tang, Y.; Meddings, J.B.; Van Eldik, L.J.; Watterson, D.M.; Clarke, L.L.; Mrsny, R.J.; Turner, J. Epithelial Myosin Light Chain Kinase–Dependent Barrier Dysfunction Mediates T Cell Activation–Induced Diarrhea in Vivo. *Clin. Investig.* **2005**, *115*, 2702–2715. [CrossRef]
20. Nalle, S.C.; Zuo, L.; Ong, M.L.D.M.; Singh, G.; Worthylake, A.M.; Choi, W.; Manresa, M.C.; Southworth, A.P.; Edelblum, K.L.; Baker, G.J.; et al. Graft-versus-Host Disease Propagation Depends on Increased Intestinal Epithelial Tight Junction Permeability. *Clin. Investig.* **2019**, *129*, 902–914. [CrossRef]
21. Li, C.; Chen, J.; Li, S.C. Understanding Horizontal Gene Transfer Network in Human Gut Microbiota. *Gut Pathog.* **2020**, *12*, 3 [CrossRef]
22. Sadowska-Klasa, A.; Piekarska, A.; Prejzner, W.; Bieniaszewska, M.; Hellmann, A. Colonization with Multidrug-Resistant Bacteria Increases the Risk of Complications and a Fatal Outcome after Allogeneic Hematopoietic Cell Transplantation. *Ann. Hematol.* **2018**, *97*, 509–517. [CrossRef] [PubMed]
23. Sears, C.L. Molecular Physiology and Pathophysiology of Tight Junctions V. Assault of the Tight Junction by Enteric Pathogens. *Am. J. Physiol. Gastrointest. Liver Physiol.* **2000**, *279*, G1129–G1134. [CrossRef] [PubMed]
24. Fredricks, D.N. The Gut Microbiota and Graft-versus-Host Disease. *J. Clin. Investig.* **2019**, *129*, 1808. [CrossRef] [PubMed]
25. Cooke, K.R.; Gerbitz, A.; Crawford, J.M.; Teshima, T.; Hill, G.R.; Tesolin, A.; Rossignol, D.P.; Ferrara, J.L.M. LPS Antagonism Reduces Graft-versus-Host Disease and Preserves Graft-versus-Leukemia Activity after Experimental Bone Marrow Transplantation. *J. Clin. Investig.* **2001**, *107*, 1581–1589. [CrossRef] [PubMed]
26. Puerta-Alcalde, P.; Chumbita, M.; Charry, P.; Castaño-Díez, S.; Cardozo, C.; Moreno-García, E.; Marco, F.; Suárez-Lledo, M.; Garcia-Pouton, N.; Morata, L.; et al. Risk Factors for Mortality in Hematopoietic Stem Cell Transplantation Recipients with Bloodstream Infection: Points To Be Addressed by Future Guidelines. *Transplant. Cell. Ther.* **2021**, *27*, 501.e1–501.e6. [CrossRef] [PubMed]
27. Perez, P.; Patiño, J.; Estacio, M.; Pino, J.; Manzi, E.; Medina, D. Bacteremia in pediatric patients with hematopoietic stem cell transplantation. *Rev. Bras. Hematol. Hemoter.* **2020**, *42*, 5–11. [CrossRef] [PubMed]
28. Romick-Rosendale, L.E.; Haslam, D.B.; Lane, A.; Denson, L.; Lake, K.; Wilkey, A.; Watanabe, M.; Bauer, S.; Litts, B.; Luebbering, N.; et al. Antibiotic Exposure and Reduced Short Chain Fatty Acid Production after Hematopoietic Stem Cell Transplant. *Biol. Blood Marrow Transplant.* **2018**, *24*, 2418–2424. [CrossRef]
29. Muratore, E.; Leardini, D.; Baccelli, F.; Venturelli, F.; Prete, A.; Masetti, R. Nutritional Modulation of the Gut Microbiome in Allogeneic Hematopoietic Stem Cell Transplantation Recipients. *Front. Nutr.* **2022**, *9*, 993668. [CrossRef]
30. Lin, D.; Hu, B.; Li, P.; Zhao, Y.; Xu, Y.; Wu, D. Roles of the Intestinal Microbiota and Microbial Metabolites in Acute GVHD. *Exp. Hematol. Oncol.* **2021**, *10*, 49. [CrossRef]

31. van der Velden, W.J.F.M.; Netea, M.G.; de Haan, A.F.J.; Huls, G.A.; Donnelly, J.P.; Blijlevens, N.M.A. Role of the Mycobiome in Human Acute Graft-versus-Host Disease. *Biol. Blood Marrow Transplant.* **2013**, *19*, 329–332. [CrossRef]
32. Malard, F.; Gasc, C.; Plantamura, E.; Doré, J. High Gastrointestinal Microbial Diversity and Clinical Outcome in Graft-versus-Host Disease Patients. *Bone Marrow Transplant.* **2018**, *53*, 1493–1497. [CrossRef] [PubMed]
33. Bolyen, E.; Rideout, J.R.; Dillon, M.R.; Bokulich, N.A.; Abnet, C.C.; Al-Ghalith, G.A.; Alexander, H.; Alm, E.J.; Arumugam, M.; Asnicar, F.; et al. Reproducible, Interactive, Scalable and Extensible Microbiome Data Science Using QIIME 2. *Nat. Biotechnol.* **2019**, *37*, 852–857. [CrossRef] [PubMed]
34. Callahan, B.J.; McMurdie, P.J.; Rosen, M.J.; Han, A.W.; Johnson, A.J.A.; Holmes, S.P. DADA2: High-Resolution Sample Inference from Illumina Amplicon Data. *Nat. Methods* **2016**, *13*, 581–583. [CrossRef] [PubMed]
35. Robeson, M.S.; O'Rourke, D.R.; Kaehler, B.D.; Ziemski, M.; Dillon, M.R.; Foster, J.T.; Bokulich, N.A. RESCRIPt: Reproducible Sequence Taxonomy Reference Database Management for the Masses. *bioRxiv* **2020**, *17*, e1009581. [CrossRef]
36. Bokulich, N.A.; Kaehler, B.D.; Rideout, J.R.; Dillon, M.; Bolyen, E.; Knight, R.; Huttley, G.A.; Gregory Caporaso, J. Optimizing Taxonomic Classification of Marker-Gene Amplicon Sequences with QIIME 2's Q2-Feature-Classifier Plugin. *Microbiome* **2018**, *6*, 90. [CrossRef]
37. Afgan, E.; Baker, D.; Batut, B.; van den Beek, M.; Bouvier, D.; Cech, M.; Chilton, J.; Clements, D.; Coraor, N.; Grüning, B.A.; et al. The Galaxy Platform for Accessible, Reproducible and Collaborative Biomedical Analyses: 2018 Update. *Nucleic Acids Res.* **2018**, *46*, W537–W544. [CrossRef]
38. Babraham Bioinformatics—FastQC a Quality Control Tool for High Throughput Sequence Data. Available online: https://www.bioinformatics.babraham.ac.uk/projects/fastqc/ (accessed on 14 July 2023).
39. Bolger, A.M.; Lohse, M.; Usadel, B. Trimmomatic: A Flexible Trimmer for Illumina Sequence Data. *Bioinformatics* **2014**, *30*, 2114–2120. [CrossRef]
40. Nurk, S.; Meleshko, D.; Korobeynikov, A.; Pevzner, P.A. MetaSPAdes: A New Versatile Metagenomic Assembler. *Genome Res.* **2017**, *27*, 824–834. [CrossRef]
41. Lu, J.; Rincon, N.; Wood, D.E.; Breitwieser, F.P.; Pockrandt, C.; Langmead, B.; Salzberg, S.L.; Steinegger, M. Metagenome Analysis Using the Kraken Software Suite. *Nat. Protoc.* **2022**, *17*, 2815–2839. [CrossRef]
42. Bortolaia, V.; Kaas, R.S.; Ruppe, E.; Roberts, M.C.; Schwarz, S.; Cattoir, V.; Philippon, A.; Allesoe, R.L.; Rebelo, A.R.; Florensa, A.F.; et al. ResFinder 4.0 for Predictions of Phenotypes from Genotypes. *J. Antimicrob. Chemother.* **2020**, *75*, 3491–3500. [CrossRef]
43. Sayers, E.W.; Bolton, E.E.; Brister, J.R.; Canese, K.; Chan, J.; Comeau, D.C.; Connor, R.; Funk, K.; Kelly, C.; Kim, S.; et al. Database Resources of the National Center for Biotechnology Information. *Nucleic Acids Res.* **2022**, *50*, D20–D26. [CrossRef] [PubMed]
44. Segata, N.; Izard, J.; Waldron, L.; Gevers, D.; Miropolsky, L.; Garrett, W.S.; Huttenhower, C. Metagenomic Biomarker Discovery and Explanation. *Genome Biol.* **2011**, *12*, R60. [CrossRef] [PubMed]
45. Dhariwal, A.; Chong, J.; Habib, S.; King, I.L.; Agellon, L.B.; Xia, J. MicrobiomeAnalyst: A Web-Based Tool for Comprehensive Statistical, Visual and Meta-Analysis of Microbiome Data. *Nucleic Acids Res.* **2017**, *45*, W180–W188. [CrossRef] [PubMed]

antibiotics

MDPI

Article

In Vitro Susceptibility of Aztreonam-Vaborbactam, Aztreonam-Relebactam and Aztreonam-Avibactam Associations against Metallo-β-Lactamase-Producing Gram-Negative Bacteria

Cécile Emeraud [1,2], Sandrine Bernabeu [1,2] and Laurent Dortet [1,2,*]

1 Department of Bacteriology-Hygiene, Bicêtre Hospital, Assistance Publique—Hôpitaux de Paris, 94270 Le Kremlin-Bicêtre, France; cecile.emeraud@aphp.fr (C.E.); sandrine.bernabeu@aphp.fr (S.B.)

2 INSERM UMR 1184, RESIST Unit, Faculty of Medicine, Paris-Saclay University, 94270 Le Kremlin-Bicêtre, France

* Correspondence: laurent.dortet@aphp.fr; Tel.: +33-145213624

Abstract: Background: Despite the availability of new options (ceftazidime-avibactam, imipenem-relebactam, meropenem-vaborbactam and cefiderocol), it is still very difficult to treat infections caused by metallo-β-lactamase (MBLs)-producers resistant to aztreonam. The in vitro efficacy of aztreonam in association with avibactam, vaborbactam or relebactam was evaluated on a collection of MBL-producing Enterobacterales, MBL-producing *P. aeruginosa* and highly drug-resistant *S. maltophilia*. Methods: A total of fifty-two non-duplicate MBL-producing Enterobacterales, five MBL-producing *P. aeruginosa* and five multidrug-resistant *S. maltophila* isolates were used in this study. The minimum inhibitory concentrations (MICs) of aztreonam, meropenem-vaborbactam and imipenem-relebactam were determined by Etest® (bioMérieux, La Balme-les-Grottes) according to EUCAST recommendations. For aztreonam-avibactam, aztreonam-vaborbactam and aztreonam-relebactam associations, the MICs were determined using Etest® on Mueller-Hinton (MH) agar supplemented with 8 mg/L of avibactam, 8 mg/L of vaborbactam and 4 mg/L of relebactam. The MICs were interpreted according to EUCAST guidelines. Results: The susceptibility rates of aztreonam-avibactam, aztreonam-vaborbactam and aztreonam-relebactam with a standard exposure of aztreonam (1g × 3, IV) were 84.6% (44/52), 55.8% and 34.6% for Enterobacterales and 0% for all combinations for *P. aeruginosa* and *S. maltophila*. The susceptibility rates of aztreonam-avibactam, aztreonam-vaborbactam and aztreonam-relebactam with a high exposure of aztreonam (2g × 4, IV) were 92.3%, 78.9% and 57.7% for Enterobacterales, 75%, 60% and 60% for *P. aeruginosa* and 100%, 100% and 40% for *S. maltophila*. Conclusions: As previously demonstrated for an aztreonam/ceftazidime-avibactam combination, aztreonam plus imipenem-relebactam and aztreonam plus meropenem-vaborbactam might be useful options, but with potentially lower efficiency, to treat infections caused by aztreonam-non-susceptible MBL-producing Gram-negative strains.

Keywords: metallo-β-lactamase; aztreonam; β-lactamase inhibitors

1. Introduction

The extensive dissemination of carbapenemase-producing Gram-negative bacteria poses a global threat to public health. It is crucial to implement new therapeutic strategies to treat infections caused by highly resistant pathogens. Carbapenemase belong to three of four classes of the Ambler classification: class A carbapenemases (mostly KPC types) [1], class B carbapenemases or metallo-β-lactamases (MBLs) (mostly NDM, VIM, or IMP types) [2], and class D carbapenemases (mostly OXA-48-like types in Enterobacterales) [3]. Recently, new therapeutic alternatives have been marked for the treatment of carbapenemase producers. These new drugs include ceftazidime-avibactam, meropenem-vaborbactam, imipenem-relebactam and cefiderocol. Despite the availability of these new options, it is still very difficult to treat infections caused by MBL producers resistant to aztreonam. Indeed, these isolates are resistant to ceftazidime-avibactam, meropenem-vaborbactam and imipenem-relebactam through the MBL-hydrolyzing activity, which is not inhibited by the added β-lactamase inhibitor. Regarding cefiderocol, it might be active on MBLs, but NDM producers have cefiderocol's MIC_{50} close to the clinical breakpoint of 2 mg/L, resulting in only 70% susceptibility [4]. Since current MBLs, such as VIM, NDM or IMP, are not able to hydrolyze aztreonam, resistance to this antimicrobial in MBL producers is due to the co-production of an ESBL or an AmpC. Accordingly, it has been demonstrated that the concomitant use of aztreonam plus ceftazidime-avibactam or amoxicillin-clavulanate or ceftolozane-tazobactam are accurate options to treat infections caused by MBL + ESBL or MBL + AmpC producers, with aztreonam plus ceftazidime-avibactam being the most powerful strategy [5]. Unfortunately, is has been recently shown that aztreonam-avibactam resistance emerged in NDM-producing *E. coli* through the association of modified penicillin-binding protein 3 (PBP3) and the production of CMY-42 cephalosporinase [6]. Thus, it is crucial to consider other therapeutic alternatives. In this study, we evaluated the susceptibility of aztreonam-resistant MBL-producing isolates to aztreonam in association with two other marketed β-lactamase inhibitors: vaborbactam and relebactam. Vaborbactam, a cyclic boronic β-lactamase inhibitor and relebactam, a diazabicyclooctane inhibitor, are very potent against KPC enzymes, but are also active on diverse class A (including ESBL) and class C β-lactamases (AmpC) [7]. As previously demonstrated for aztreonam-avibactam [5], aztreonam-vaborbactam and aztreonam-relebactam seemed to be effective on NDM-producing *K. pneumoniae* [8] and few multidrug-resistant *S. maltophila* [9]. Here, we assessed the susceptibility of aztreonam in combination with vaborbactam or relebactam or avibactam on a collection of MBL producers including Enterobacterales (including species other than *K. pneumoniae*), *P. aeruginosa* and *S. maltophilia*.

2. Results

A total of fifty-two non-duplicate MBL-producing Enterobacterales, five MBL-producing *P. aeruginosa* and five multidrug-resistant *S. maltophila* isolates were used in this study. These isolates have already been tested for aztreonam-avibactam in a previous study [5]. All strains were resistant to aztreonam due to the co-production of an ESBL and/or a cephalosporinase. The MBL-producing Enterobacterales included 30 NDM-producers with 11 isolates co-producing one OXA-48-like carbapenemase, fourteen VIM-producers, six IMP-producers, one GIM-1-producer and one TMB-1-producer. The MBL-producing *P. aeruginosa* were one VIM-2-, three IMP-1- and one IMP-2-producers. *S. maltophila* isolates were resistant to all β-lactams including ticarcillin-clavulanate, and to all other antimicrobials including fluoroquinolones and trimethoprim-sulfamethoxazole [5].

None of the strains included in the study were sensitive to aztreonam at standard exposure (the MICs range from 2 to >256 mg/L). Regarding the MIC distribution of each association, aztreonam-relebactam appeared to be more efficient than aztreonam-vaborbactam, but both are less efficient than aztreonam-avibactam (Figure 1).

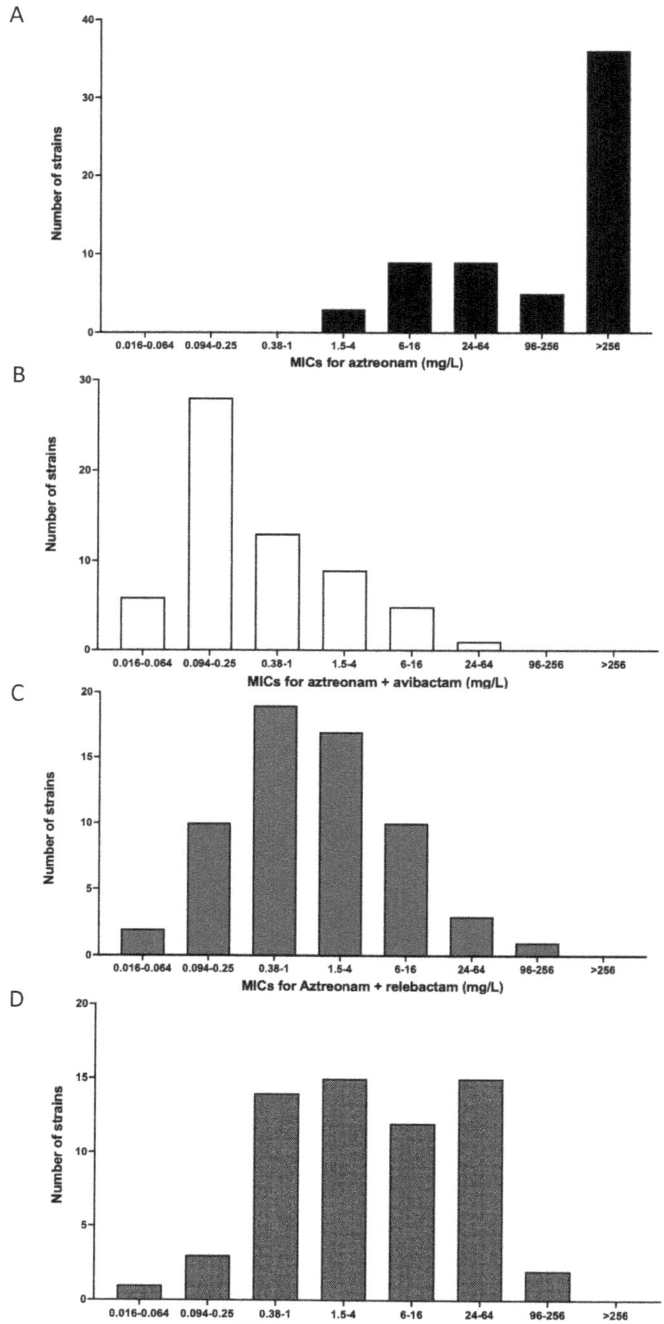

Figure 1. MICs of aztreonam (**A**), aztreonam-avibactam (**B**), aztreonam-relebactam (**C**) and aztreonam-vaborbactam (**D**).

The addition of avibactam, relebactam or vaborbactam decreases the MIC of aztreonam by at least four dilutions for 100%, 93.5% and 90.3% of the strains, respectively (Table 1). The susceptibility rates of aztreonam-avibactam, aztreonam-vaborbactam and aztreonam-relebactam with a standard exposure of aztreonam (meaning 1g × 3, IV according to EUCAST guidelines) (≤1 mg/L for Enterobacterales and ≤0.001 mg/L for *S. maltophila* and *P. aeruginosa*) were 84.6% (44/52), 55.8% (29/52) and 34.6% (18/52) for Enterobacterales and 0% for all combinations for *P. aeruginosa* and *S. maltophila*, respectively. The susceptibility rates of aztreonam-avibactam, aztreonam-vaborbactam and aztreonam-relebactam with a high exposure of aztreonam (meaning 2g × 4, IV according to EUCAST guidelines) (≤4 mg/L for Enterobacterales and ≤16 mg/L for *S. maltophila* and *P. aeruginosa*) were 92.3% (48/52), 78.9% (41/52) and 57.7% (30/52) for Enterobacterales, 75% (4/5), 60% (3/5) and 60% (3/5) for *P. aeruginosa* and 100% (5/5), 100% (5/5) and 40% (2/5) for *S. maltophila* isolates, respectively (Table 1). As previously observed for the aztreonam-avibactam combination [5], the MICs of aztreonam-relebactam and aztreonam-vaborbactam were higher for NDM-producing *E. coli* compared to NDM-producing *K. pneumoniae*.

Table 1. MICs of aztreonam (ATM), aztreonam-avibactam (ATM + AVI), aztreonam-relebactam (ATM + REL) and aztreonam-vaborbactam (ATM + VAB).

Species	β-Lactamases	MICs (mg/L)			
		ATM	ATM + AVI	AZM+ REL	ATM + VAB
E. coli ATCC 53126		0.047	ND	0.047	0.047
K. pneumoniae ATCC 700603		64	ND	0.5	2
E. coli	KPC	>256	ND	0.38	2
K. pneumoniae	KPC	>256	ND	0.125	0.25
E. coli	NDM-1 + OXA-1 + OXA-10 + CMY-16 + TEM-1	32	0.125	0.125	0.5
E. coli	NDM-1 + CTX-M-15 + TEM-1	>256	1	12	24
E. coli	NDM-1 + OXA-1 + OXA-2 + CTX-M-15 + TEM-1	>256	2	12	32
E. coli	NDM-1 + CTX-M-15 + TEM-1	>256	6	32	192
E. coli	NDM-4 + CTX-M-15 + OXA-1	>256	6	16	24
E. coli	NDM-4 + CTX-M-15 + CMY-6	>256	6	8	24
E. coli	NDM-5 + TEM-1 + CTX-M-15	>256	8	24	64
E. coli	NDM-6 + CTX-M-15 + OXA-1	>256	1	3	8
E. coli	NDM-7 + CTX-M-15	>256	4	12	32
K. pneumoniae	NDM-1 + CTX-M-15 + SHV-11 + OXA-1	>256	0.125	1	1.5
K. pneumoniae	NDM-1 + CTX-M-15 + CMY-4 + OXA-1	>256	0.75	4	48
K. pneumoniae	NDM-1 + CTX-M-15 + OXA-1 + OXA-9 + TEM-1 + SHV-28 + SHV-11	>256	0.25	4	12
K. pneumoniae	NDM-1 + OXA-1 + SHV-11	>256	0.047	0.094	0.094
K. pneumoniae	NDM-1 + OXA-1 + CTX-M-15 + TEM-1 + SHV-28 + OXA-9 + CMY-6	>256	0.047	0.75	0.75
K. pneumoniae	NDM-1 + TEM-1 + CTX-M-15 + SHV-12 + OXA-9	>256	0.125	1.5	1.5
K. pneumoniae	NDM-1 + TEM-1 + CTX-M-15 + SHV-12 + OXA-9	>256	0.125	1.5	12
K. pneumoniae	NDM-1 + TEM-1 + CTX-M-15 + SHV-11 + OXA-1	>256	0.064	0.75	0.75
P. stuartii	NDM-1 + OXA-1 + CMY-6 + TEM-1	8	0.032	0.016	0.032
Salmonella enterica	NDM-1 + CTX-M-15 + TEM-1 + OXA-1 + OXA-9 + OXA-10	>256	0.125	0.75	0.75
E. coli	VIM-1 + CTX-M-3	>256	0.125	1.5	12
E. coli	VIM-4 + CTX-M15	16	1.5	1.5	8
K. pneumoniae	VIM-1 + SHV-5	>256	0.25	3	24
K. pneumoniae	VIM-1 + SHV-12	>256	0.125	1	6
K. pneumoniae	VIM-1 + CTX-M-15	>256	0.19	0.125	0.125
K. pneumoniae	VIM-1 + SHV-5	16	0.19	1	1
K. pneumoniae	VIM-1 + TEM-1 + SHV-5	>256	0.25	1	6

Table 1. *Cont.*

Species	β-Lactamases	MICs (mg/L)			
		ATM	ATM + AVI	AZM+ REL	ATM + VAB
K. pneumoniae	VIM-1 + SHV-5	>256	0.25	12	32
K. pneumoniae	VIM-1 + SHV-5	>256	0.125	1.5	8
K. pneumoniae	VIM-19 + CTX-M-3 + TEM-1 + SHV-1	6	0.047	0.032	0.094
E. cloacae	VIM-1 + SHV-70	256	0.094	0.5	2
E. cloacae	VIM-4 + CTX-M-15 + TEM-1 + SHV-31	64	1	0.5	4
C. freundii	VIM-2 + TEM-1 + CTX-M-15	16	0.25	3	1.5
C. freundii	VIM-2 + TEM-1 + OXA-9 + OXA-10	32	1.5	1.5	4
E. coli	IMP-8 + SHV -12	128	0.19	0.75	4
K. pneumoniae	IMP-1 + TEM-15	3	0.094	0.125	0.38
K. pneumoniae	IMP-1 + TEM-15 + CTX-M-15	2	0.094	0.125	0.38
K. pneumoniae	IMP-8 + SHV -12	>256	0.094	0.125	4
E. cloacae	IMP-8 + SHV-12	12	0.032	0.125	0.38
S. marscecens	IMP-11	4	0.5	0.125	0.38
E. cloacae	GIM-1 + CTX-M-15	12	0.5	0.19	1
C. freundii	TMB-1 + overexpressed cephalosporinase	64	0.125	0.19	0.75
K. pneumoniae	NDM-1 + OXA-181 + SHV-11 + TEM-1 + CTX-M-15 + OXA-1	64	0.094	0.5	0.5
K. pneumoniae	NDM-1 + OXA-181 + SHV-27 + CTX-M-15 + TEM-1 + OXA-1	128	0.25	1	1
K. pneumoniae	NDM-1 + OXA-181 + SHV-11 + CTX-M-15 + OXA-1	256	0.19	1	4
K. pneumoniae	NDM-1 + OXA-181 + SHV-11 + TEM-1 + CTX-M-15 + OXA-9	>256	0.19	0.75	16
K. pneumoniae	NDM-1 + OXA-181 + SHV-2 + CTX-M-15 + OXA-1	>256	0.125	0.75	1
C. freundii	NDM-1 + OXA-181 + OXA-1 + OXA-9 + OXA-10 + CTX-M-15 + TEM-1	>256	0.75	4	32
E. coli	NDM-1 + OXA-48 + CTX-M-15	32	0.094	0.5	1.5
E. coli	NDM-1 + OXA-48 + CTX-M-15	>256	0.75	6	16
K. pneumoniae	NDM-1 + OXA-232 + CTX-M-15	64	0.094	0.5	1.5
E. coli	NDM-1 + OXA-232 + CTX-M-15	>256	1	6	1.5
E. coli	NDM-5 + OXA-232 + CTX-M-15	>256	1	6	32
Percentage of susceptible strains of Enterobacterales with standard exposure		**0%**	**84.6%**	**55.8%**	**34.6%**
Percentage of susceptible strains of Enterobacterales with high exposure		**8.1%**	**92.3%**	**78.9%**	**57.7%**
Percentage of resistant strains of Enterobacterales		**91.9%**	**13.5%**	**21.1%**	**42.3%**
S. maltophilia		>256	0.5	1.5	1.5
S. maltophilia		>256	2	3	64
S. maltophilia		>256	0.5	1	4
S. maltophilia		>256	2	3	48
S. maltophilia		>256	1	1	32
P. aeruginosa	VIM-2 + overexpressed Cephalosporinase	32	8	32	24
P. aeruginosa	IMP-2 + overexpressed Cephalosporinase	6	1.5	2	2
P. aeruginosa	IMP-1 + overexpressed Cephalosporinase	24	3	4	16
P. aeruginosa	IMP-1 + overexpressed Cephalosporinase	96	32	128	96
P. aeruginosa	IMP-1 + overexpressed Cephalosporinase	12	3	8	6
Percentage of susceptible strains with standard exposure of aztreonam in the combination (total)		**0%**	**66.1%**	**45.2%**	**29.0%**
Percentage of susceptible strains with high exposure of aztreonam in the combination (total)		**4.8%**	**87.1%**	**79.1%**	**56.4%**
Percentage of resistant strains (total)		**95.2%**	**12.9%**	**21.0%**	**43.6%**
Percentage of strains with aztreonam MIC reduction ≥ 4-fold dilution			**100%**	**93.5%**	**90.3%**

Black: resistant; Grey: susceptible at high exposure; Blank: susceptible at standard dosage.

3. Discussion

In conclusion, as previously demonstrated for the aztreonam/ceftazidime-avibactam combination [10–18], aztreonam plus imipenem-relebactam [19–21] and aztreonam plus meropenem-vaborbactam [21] might be useful options to treat infections caused by aztreonam-non-susceptible MBL-producing Gram-negative strains. Despite the fact that avibactam seems to be the most potent inhibitor in association with aztreonam, it should be noted that ceftazidime is strongly hydrolyzed by ESBL and MBL with MICs >256 mg/L resulting in a non-residual effect of this drug in the tripartite combination aztreonam-ceftazidime-avibactam. Contrarily, imipenem and meropenem are not hydrolyzed by ESBL or AmpC and might retain partial activity on MBL (sometimes the MICs range from 2 to 8 mg/L). Accordingly, the tripartite associations of aztreonam-imipenem-relebactam and aztreonam-meropenem-vaborbactam might benefit not only the activity of the inhibitor with aztreonam, but also partly the residual activity of the carbapenem. Notably, our collection of Gram-negative bacteria included only a few isolates of *P. aeruginosa* and *S. maltophilia*. Due to this limitation, the results of the susceptibility of these two species have to be confirmed in further studies.

According to the recent guidelines of the Infectious Diseases Society of America (IDSA) [22] and the European Society of Clinical Microbiology and Infectious Diseases (ESCMID) [23], aztreonam-avibactam association produced by injecting aztreonam and ceftazidime-avibactam simultaneously [24] is considered the first-line treatment for infections caused by MBL producers. This is particularly the case for NDM producers that often display an increased MIC to cefiderocol [4,25–27], the other potential option for the treatment of infections caused by MBL producers [22,23].

In our collection, four NDM-producing *E. coli* displayed resistance to the aztreonam-avibactam association. In *E. coli*, the impact of the alteration of PBP3 has been recently described to be responsible for increased resistance to aztreonam-avibactam [11] and a cross-resistance to cefiderocol [28,29]. With the recent emergence of aztreonam-avibactam resistance among carbapenemase-producing Enterobacterales [11,29,30], these tripartite associations of aztreonam-imipenem-relebactam and aztreonam-meropenem-vaborbactam have to be taken into consideration as last-resort treatment options. Unfortunately, resistance to aztreonam-meropenem-vaborbactam has already been reported in a KPC variant producing *K. pneumoniae* [31]. Accordingly, the rapid development of additional options, such as new combinations with more potent inhibitors (e.g., zidebactam and taniborbactam) are mandatory [32–34]. However, as it was observed for aztreonam-avibactam, which is not already marketed, resistance to these very novel inhibitors already exists, as reported for cefepime-taniborbactam and NDM-9 producers [35,36].

In conclusion, aztreonam-vaborbactam and aztreonam-relebactam could be useful options for the treatment of infections caused by aztreonam-resistant MBL-producing isolates but with a potential lower efficiency compared to aztreonam-avibactam.

4. Materials and Methods

A total of 52 non-duplicate MBL-producing Enterobacterales, 5 MBL-producing *P. aeruginosa* and 5 multidrug-resistant *S. maltophila* isolates were used in this study. All strains were sequenced using the Illumina technique and the resistance genes were identified using Resfinder 4.1 (https://cge.food.dtu.dk/services/ResFinder/, 27 September 2023).

The MICs of aztreonam, meropenem-vaborbactam and imipenem-relebactam were determined by Etest® (bioMérieux, La Balme-les-Grottes) according to EUCAST recommendations. For aztreonam-avibactam, aztreonam-vaborbactam and aztreonam-relebactam associations, MICs were determined using Etest® on MH agar supplemented with 8 mg/L of avibactam (provided by Pfizer, France), 8 mg/L of vaborbactam (CliniSciences, Nanterre, France) and 4 mg/L of relebactam (provided by MSD, Puteaux, France). The MICs of aztreonam-avibactam, aztreonam-vaborbactam and aztreonam-relebactam were interpreted in the same way as aztreonam alone, according to EUCAST guidelines as updated in 2022 (Table 2).

Table 2. EUCAST breakpoint of aztreonam for Enterobacterales and *Pseudomonas aeruginosa* (https://www.eucast.org/clinical_breakpoints, accessed on 27 September 2023).

	MICs (mg/L)	
	Enterobacterales	***P. aeruginosa***
Susceptible with standard exposure	≤1	≤0.001
Susceptible with high exposure	≤4	≤16
Resistant	>4	>16

Since no recommendation exists regarding these molecules for *S. maltophila*, interpretation criteria of *Pseudomonas* spp. were used (Table 3). A control of the accuracy of β-lactam inhibitor-supplemented MH agar was performed. For this purpose, the MICs of 13 strains, obtained with the imipenem and meropenem Etest® (bioMérieux, France) on MH agar supplemented with 8 mg/L of vaborbactam and 4 mg/L of relebactam, were compared to the MICs obtained with the imipenem-relebactam and meropenem-vaborbactam Etest® (bioMérieux, France). The difference in MICs obtained between the two methods did not exceed two dilutions, validating the methodology used in this study (Table 3). However, due to the relatively low number of tested isolates, the full validity of this method might have to be confirmed in further studies.

Table 3. MICs of imipenem-relebactam (IMP + REL) and meropenem-vaborbactam (MEM + VAB) obtained using Etest® and supplemented Mueller–Hinton agar.

		MICs			
		IMP+ REL Etest®	**IMP+ REL Agar Method [a]**	**MEM+VAB Etest®**	**MEM+VAB Agar Method [b]**
E. coli	NDM-1 + OXA-1 + OXA-2 + CTX-M-15 + TEM-1	4	8	6	8
E. coli	NDM-4 + CTX-M-15 + OXA-1	16	>32	32	>32
E. coli	NDM-5 + TEM-1 + CTX-M-15	>32	>32	>32	>32
K. pneumoniae	NDM-1 + TEM-1 + CTX-M-15 + SHV-12 + OXA-9	2	1	2	4
P. stuartii	NDM-1 + OXA-1 + CMY-6 + TEM-1	4	8	2	2
E. coli	VIM-4 + ESBL	6	4	3	1
K. pneumoniae	VIM-1 + SHV-5	3	4	1.5	2
K. pneumoniae	VIM-19 + CTX-M-3 + TEM-1 + SHV-1	1	2	0.75	1
K. pneumoniae	NDM-1 + OXA-181 + SHV-11 + CTX-M-15 + OXA-1	4	8	12	12
E. coli	NDM-1 + OXA-48 + ESBL	2	4	2	2
E. coli	NDM-5 + OXA-232 + ESBL	2	2	16	8
S. maltophilia		>32	>32	>32	>32
P. aeruginosa	IMP-1 + overexpressed Cephalosporinase	>32	16	>32	>32

[a] Mueller–Hinton agar supplemented with 4 mg/L of relebactam. [b] Mueller–Hinton agar supplemented with 8 mg/L of vaborbactam.

Of note is the fact that the accuracy of determining the MIC of aztreonam-avibactam using MH agar supplemented with 8 mg/L of avibactam has already been demonstrated on this strain collection. In the previous study, this method was also demonstrated to give similar results with the Etest® strip superposition method [5]. *E. coli* ATCC 53126, *K. pneumoniae* ATCC 700,603 and two KPC-producing strains were used as quality controls.

Author Contributions: Conceptualization, L.D.; methodology, L.D. and C.E.; validation, L.D. and C.E.; formal analysis, L.D. and C.E.; investigation, S.B. and C.E.; resources, L.D.; data curation, C.E.; writing—original draft preparation, C.E.; writing—review and editing, L.D.; supervision, L.D. All authors have read and agreed to the published version of the manuscript.

Funding: This research received no external funding.

Institutional Review Board Statement: Not applicable.

Informed Consent Statement: Not applicable.

Data Availability Statement: The data presented in this study are available on request from the corresponding author.

Conflicts of Interest: The authors declare no conflict of interest.

References

1. Naas, T.; Dortet, L.; Iorga, B.I. Structural and Functional Aspects of Class A Carbapenemases. *Curr. Drug Targets* **2016**, *17*, 1006–1028. [CrossRef] [PubMed]
2. Mojica, M.F.; Bonomo, R.A.; Fast, W. B1-Metallo-beta-Lactamases: Where Do We Stand? *Curr. Drug Targets* **2016**, *17*, 1029–1050. [CrossRef] [PubMed]
3. Poirel, L.; Potron, A.; Nordmann, P. OXA-48-like carbapenemases: The phantom menace. *J. Antimicrob. Chemother.* **2012**, *67*, 1597–1606. [CrossRef] [PubMed]
4. Bonnin, R.A.; Emeraud, C.; Jousset, A.B.; Naas, T.; Dortet, L. Comparison of disk diffusion, MIC test strip and broth microdilution methods for cefiderocol susceptibility testing on carbapenem-resistant enterobacterales. *Clin. Microbiol. Infect.* **2022**, *28*, 1156.e1–1156.e5. [CrossRef]
5. Emeraud, C.; Escaut, L.; Boucly, A.; Fortineau, N.; Bonnin, R.A.; Naas, T.; Dortet, L. Aztreonam plus Clavulanate, Tazobactam, or Avibactam for Treatment of Infections Caused by Metallo-beta-Lactamase-Producing Gram-Negative Bacteria. *Antimicrob. Agents Chemother.* **2019**, *63*. [CrossRef] [PubMed]
6. Sadek, M.; Juhas, M.; Poirel, L.; Nordmann, P. Genetic Features Leading to Reduced Susceptibility to Aztreonam-Avibactam among Metallo-beta-Lactamase-Producing *Escherichia coli* Isolates. *Antimicrob. Agents Chemother.* **2020**, *64*. [CrossRef]
7. Zhanel, G.G.; Lawrence, C.K.; Adam, H.; Schweizer, F.; Zelenitsky, S.; Zhanel, M.; Lagace-Wiens, P.R.S.; Walkty, A.; Denisuik, A.; Golden, A.; et al. Imipenem-Relebactam and Meropenem-Vaborbactam: Two Novel Carbapenem-beta-Lactamase Inhibitor Combinations. *Drugs* **2018**, *78*, 65–98. [CrossRef]
8. Maraki, S.; Mavromanolaki, V.E.; Moraitis, P.; Stafylaki, D.; Kasimati, A.; Magkafouraki, E.; Scoulica, E. Ceftazidime-avibactam, meropenen-vaborbactam, and imipenem-relebactam in combination with aztreonam against multidrug-resistant, metallo-beta-lactamase-producing *Klebsiella pneumoniae*. *Eur. J. Clin. Microbiol. Infect. Dis.* **2021**, *40*, 1755–1759. [CrossRef]
9. Biagi, M.; Lamm, D.; Meyer, K.; Vialichka, A.; Jurkovic, M.; Patel, S.; Mendes, R.E.; Bulman, Z.P.; Wenzler, E. Activity of Aztreonam in Combination with Avibactam, Clavulanate, Relebactam, and Vaborbactam against Multidrug-Resistant *Stenotrophomonas maltophilia*. *Antimicrob. Agents Chemother.* **2020**, *64*. [CrossRef]
10. Kang, Y.; Xie, L.; Yang, J.; Cui, J. Optimal treatment of ceftazidime-avibactam and aztreonam-avibactam against bloodstream infections or lower respiratory tract infections caused by extensively drug-resistant or pan drug-resistant (XDR/PDR) *Pseudomonas aeruginosa*. *Front. Cell Infect. Microbiol.* **2023**, *13*, 1023948. [CrossRef]
11. Livermore, D.M.; Mushtaq, S.; Vickers, A.; Woodford, N. Activity of aztreonam/avibactam against metallo-beta-lactamase-producing Enterobacterales from the UK: Impact of penicillin-binding protein-3 inserts and CMY-42 beta-lactamase in *Escherichia coli*. *Int. J. Antimicrob. Agents* **2023**, *61*, 106776. [CrossRef] [PubMed]
12. Rossolini, G.M.; Stone, G.; Kantecki, M.; Arhin, F.F. In vitro activity of aztreonam/avibactam against isolates of Enterobacterales collected globally from ATLAS in 2019. *J. Glob. Antimicrob. Resist.* **2022**, *30*, 214–221. [CrossRef] [PubMed]
13. Sader, H.S.; Castanheira, M.; Kimbrough, J.H.; Kantro, V.; Mendes, R.E. Aztreonam/avibactam activity against a large collection of carbapenem-resistant Enterobacterales (CRE) collected in hospitals from Europe, Asia and Latin America (2019–21). *JAC Antimicrob. Resist.* **2023**, *5*, dlad032. [CrossRef] [PubMed]
14. Sonnevend, A.; Ghazawi, A.; Darwish, D.; Barathan, G.; Hashmey, R.; Ashraf, T.; Rizvi, T.A.; Pal, T. In vitro efficacy of ceftazidime-avibactam, aztreonam-avibactam and other rescue antibiotics against carbapenem-resistant Enterobacterales from the Arabian Peninsula. *Int. J. Infect. Dis.* **2020**, *99*, 253–259. [CrossRef]
15. Wang, X.; Zhang, F.; Zhao, C.; Wang, Z.; Nichols, W.W.; Testa, R.; Li, H.; Chen, H.; He, W.; Wang, Q.; et al. In vitro activities of ceftazidime-avibactam and aztreonam-avibactam against 372 Gram-negative bacilli collected in 2011 and 2012 from 11 teaching hospitals in China. *Antimicrob. Agents Chemother.* **2014**, *58*, 1774–1778. [CrossRef]
16. Wise, M.G.; Karlowsky, J.A.; Mohamed, N.; Kamat, S.; Sahm, D.F. In vitro activity of aztreonam-avibactam against Enterobacterales isolates collected in Latin America, Africa/Middle East, Asia, and Eurasia for the ATLAS Global Surveillance Program in 2019–2021. *Eur. J. Clin. Microbiol. Infect. Dis.* **2023**, *42*, 1135–1143. [CrossRef]
17. Zhang, B.; Zhu, Z.; Jia, W.; Qu, F.; Huang, B.; Shan, B.; Yu, H.; Tang, Y.; Chen, L.; Du, H. In vitro activity of aztreonam-avibactam against metallo-beta-lactamase-producing Enterobacteriaceae-A multicenter study in China. *Int. J. Infect. Dis.* **2020**, *97*, 11–18. [CrossRef]
18. Zou, C.; Wei, J.; Shan, B.; Chen, X.; Wang, D.; Niu, S. In vitro Activity of Ceftazidime-Avibactam and Aztreonam-Avibactam Against Carbapenem-resistant Enterobacteriaceae Isolates Collected from Three Secondary Hospitals in Southwest China Between 2018 and 2019. *Infect. Drug Resist.* **2020**, *13*, 3563–3568. [CrossRef]

19. Biagi, M.; Lee, M.; Wu, T.; Shajee, A.; Patel, S.; Deshpande, L.M.; Mendes, R.E.; Wenzler, E. Aztreonam in combination with imipenem-relebactam against clinical and isogenic strains of serine and metallo-beta-lactamase-producing enterobacterales. *Diagn. Microbiol. Infect. Dis.* **2022**, *103*, 115674. [CrossRef]
20. O'Donnell, J.N.; Putra, V.; Belfiore, G.M.; Maring, B.L.; Young, K.; Lodise, T.P. In vitro activity of imipenem/relebactam plus aztreonam against metallo-beta-lactamase-producing, OprD-deficient *Pseudomonas aeruginosa* with varying levels of *Pseudomonas*-derived cephalosporinase production. *Int. J. Antimicrob. Agents* **2022**, *59*, 106595. [CrossRef]
21. Vazquez-Ucha, J.C.; Alonso-Garcia, I.; Guijarro-Sanchez, P.; Lasarte-Monterrubio, C.; Alvarez-Fraga, L.; Cendon-Esteve, A.; Outeda, M.; Maceiras, R.; Pena-Escolano, A.; Martinez-Guitian, M.; et al. Activity of aztreonam in combination with novel beta-lactamase inhibitors against metallo-beta-lactamase-producing Enterobacterales from Spain. *Int. J. Antimicrob. Agents* **2023**, *61*, 106738. [CrossRef] [PubMed]
22. Infectious Diseases Society of America. IDSA 2023 Guidance on the Treatment of Antimicrobial Resistant Gram-Negative Infections. 2023. Available online: https://www.idsociety.org/globalassets/idsa/practice-guidelines/amr-guidance/1.0/idsa-amr-guidance-v3.0.pdf (accessed on 31 December 2022).
23. Paul, M.; Carrara, E.; Retamar, P.; Tangden, T.; Bitterman, R.; Bonomo, R.A.; de Waele, J.; Daikos, G.L.; Akova, M.; Harbarth, S.; et al. European Society of Clinical Microbiology and Infectious Diseases (ESCMID) guidelines for the treatment of infections caused by multidrug-resistant Gram-negative bacilli (endorsed by European society of intensive care medicine). *Clin. Microbiol. Infect.* **2022**, *28*, 521–547. [CrossRef] [PubMed]
24. Falcone, M.; Daikos, G.L.; Tiseo, G.; Bassoulis, D.; Giordano, C.; Galfo, V.; Leonildi, A.; Tagliaferri, E.; Barnini, S.; Sani, S.; et al. Efficacy of Ceftazidime-avibactam Plus Aztreonam in Patients With Bloodstream Infections Caused by Metallo-beta-lactamase-Producing Enterobacterales. *Clin. Infect. Dis.* **2021**, *72*, 1871–1878. [CrossRef]
25. Alzayer, M.; Alghoribi, M.F.; Alalwan, B.; Alreheli, A.; Aljohani, S.; Bosaeed, M.; Doumith, M. In vitro activity of cefiderocol against clinically important carbapenem non-susceptible Gram-negative bacteria from Saudi Arabia. *J. Glob. Antimicrob. Resist.* **2023**, *32*, 176–180. [CrossRef] [PubMed]
26. Kohira, N.; Hackel, M.A.; Ishioka, Y.; Kuroiwa, M.; Sahm, D.F.; Sato, T.; Maki, H.; Yamano, Y. Reduced susceptibility mechanism to cefiderocol, a siderophore cephalosporin, among clinical isolates from a global surveillance programme (SIDERO-WT-2014). *J. Glob. Antimicrob. Resist.* **2020**, *22*, 738–741. [CrossRef]
27. Oueslati, S.; Bogaerts, P.; Dortet, L.; Bernabeu, S.; Ben Lakhal, H.; Longshaw, C.; Glupczynski, Y.; Naas, T. In vitro Activity of Cefiderocol and Comparators against Carbapenem-Resistant Gram-Negative Pathogens from France and Belgium. *Antibiotics* **2022**, *11*, 1352. [CrossRef]
28. Poirel, L.; Ortiz de la Rosa, J.M.; Sakaoglu, Z.; Kusaksizoglu, A.; Sadek, M.; Nordmann, P. NDM-35-Producing ST167 *Escherichia coli* Highly Resistant to beta-Lactams Including Cefiderocol. *Antimicrob. Agents Chemother.* **2022**, *66*, e0031122. [CrossRef] [PubMed]
29. Simner, P.J.; Bergman, Y.; Conzemius, R.; Jacobs, E.; Tekle, T.; Beisken, S.; Tamma, P.D. An NDM-Producing *Escherichia coli* Clinical Isolate Exhibiting Resistance to Cefiderocol and the Combination of Ceftazidime-Avibactam and Aztreonam: Another Step Toward Pan-beta-Lactam Resistance. *Open Forum Infect. Dis.* **2023**, *10*, ofad276. [CrossRef]
30. Wu, S.; Ma, K.; Feng, Y.; Zong, Z. Resistance to aztreonam-avibactam due to a mutation of SHV-12 in Enterobacter. *Ann. Clin. Microbiol. Antimicrob.* **2023**, *22*, 49. [CrossRef]
31. Satapoomin, N.; Dulyayangkul, P.; Avison, M.B. *Klebsiella pneumoniae* Mutants Resistant to Ceftazidime-Avibactam Plus Aztreonam, Imipenem-Relebactam, Meropenem-Vaborbactam, and Cefepime-Taniborbactam. *Antimicrob. Agents Chemother.* **2022**, *66*, e0217921. [CrossRef]
32. Karvouniaris, M.; Almyroudi, M.P.; Abdul-Aziz, M.H.; Blot, S.; Paramythiotou, E.; Tsigou, E.; Koulenti, D. Novel Antimicrobial Agents for Gram-Negative Pathogens. *Antibiotics* **2023**, *12*, 761. [CrossRef] [PubMed]
33. Le Terrier, C.; Nordmann, P.; Freret, C.; Seigneur, M.; Poirel, L. Impact of Acquired Broad Spectrum beta-Lactamases on Susceptibility to Novel Combinations Made of beta-Lactams (Aztreonam, Cefepime, Meropenem, and Imipenem) and Novel beta-Lactamase Inhibitors in *Escherichia coli* and *Pseudomonas aeruginosa*. *Antimicrob. Agents Chemother.* **2023**, *67*, e0003923. [CrossRef] [PubMed]
34. Le Terrier, C.; Nordmann, P.; Sadek, M.; Poirel, L. In vitro activity of cefepime/zidebactam and cefepime/taniborbactam against aztreonam/avibactam-resistant NDM-like-producing *Escherichia coli* clinical isolates. *J. Antimicrob. Chemother.* **2023**, *78*, 1191–1194. [CrossRef] [PubMed]
35. Le Terrier, C.; Gruenig, V.; Fournier, C.; Nordmann, P.; Poirel, L. NDM-9 resistance to taniborbactam. *Lancet Infect. Dis.* **2023**, *23*, 401–402. [CrossRef] [PubMed]
36. Le Terrier, C.; Nordmann, P.; Buchs, C.; Di, D.Y.W.; Rossolini, G.M.; Stephan, R.; Castanheira, M.; Poirel, L. Wide dissemination of Gram-negative bacteria producing the taniborbactam-resistant NDM-9 variant: A One Health concern. *J. Antimicrob. Chemother.* **2023**, *78*, 2382–2384. [CrossRef]

antibiotics

Article

Carbapenem-Resistant *Klebsiella pneumoniae* in COVID-19 Era—Challenges and Solutions

Jozef Ficik [1,2], Michal Andrezál [3], Hana Drahovská [3], Miroslav Böhmer [4,5], Tomáš Szemes [3,4,5], Adriána Liptáková [2] and Lívia Slobodníková [2,*]

1 Institute of Clinical Microbiology, Central Military Hospital, 034 26 Ružomberok, Slovakia; ficikj@uvn.sk
2 Institute of Microbiology, Faculty of Medicine, Comenius University and the University Hospital in Bratislava, 811 08 Bratislava, Slovakia; adriana.liptakova@fmed.uniba.sk
3 Department of Molecular Biology, Faculty of Natural Sciences, Comenius University, 842 15 Bratislava, Slovakia; michal.andrezal@uniba.sk (M.A.); hana.drahovska@uniba.sk (H.D.); tomas.szemes@uniba.sk (T.S.)
4 Comenius University Science Park, 841 02 Bratislava, Slovakia; miroslav.bohmer@uniba.sk
5 Public Health Authority of the Slovak Republic, 826 45 Bratislava, Slovakia
* Correspondence: livia.slobodnikova@fmed.uniba.sk

Abstract: The COVID-19 era brought about new medical challenges, which, together with nosocomial bacterial infections, resulted in an enormous burden for the healthcare system. One of the most alarming nosocomial threats was carbapenem-resistant *Klebsiella pneumoniae* (CRKP). Monitoring CRKP incidence and antimicrobial resistance globally and locally is vitally important. In a retrospective study, the incidence of CRKP in the pre-COVID-19 period (2017–2019) and the COVID-19 pandemic (2020–2022) was investigated in the Central Military Hospital in Ružomberok, Slovak Republic. The relative incidence of CRKP significantly increased during the COVID-19 period—by 4.8 times, from 0.18 to 0.76%. At the same time, 47% of CRKP-positive patients also had COVID-19. Twenty-six KPC and sixty-nine NDM-producing isolates were identified. CRKPs isolated in the year 2022 were submitted to whole genome sequencing, and their susceptibility was tested to cefiderocol, ceftazidime–avibactam, imipenem–relebactam and meropenem–vaborbactam, with excellent results. KPC-producing isolates were also highly susceptible to colistin (92%). The NDM isolates revealed lower susceptibility rates, including only 57% colistin susceptibility. ST-307 prevailed in KPC and ST-11 in NDM isolates. Despite the excellent activity of new antimicrobials, rational antibiotic policy must be thoroughly followed, supported by complementary treatments and strict anti-epidemic precautions.

Keywords: carbapenem-resistant *Klebsiella pneumoniae*; KPC; NDM; COVID-19; antimicrobial resistance; whole genome sequencing

1. Introduction

The emergence and spread of carbapenem-resistant *Klebsiella pneumoniae* (CRKP) in hospitals worldwide [1,2], including the Slovak Republic [3,4], represent an urgent global public health issue, posing an enormous burden on hospital settings caring for high-risk patients [5]. Infections caused by these bacterial agents represent a significant therapeutic problem, primarily in patients in intensive care units, where bacterial superinfections increase the morbidity and mortality rates of critically ill and debilitated patients, with an overwhelming impact on hospitalisation costs [6,7]. Previous analyses show carbapenemase-producing bacteria effectively spreading during times of chaos—in war or in countries facing socio-economic and immigrant challenges [8]. A similar situation was seen in the COVID-19 era. The start of the COVID-19 pandemic brought about new healthcare challenges, with secondary bacterial infections in hospitalised patients posing additional health threats. COVID-19 patients experienced co-infection with carbapenemase producers [9]. Studies from various regions started to report increasing carbapenem resistance, dominated primarily by *Klebsiella pneumoniae* strains, occurring mainly in intensive-care units treating patients with severe COVID-19 [10–13]. The enormous increase in critically ill patients during the COVID-19 pandemic resulted in overcrowding of hospitals, especially the intensive care units, leading to an increased risk of nosocomial infections spreading. Lack of intensive care beds and mechanical ventilators also resulted in inter-hospital transfers of patients, with an additional risk of nosocomial agents dissemination as the pandemic impacted control and surveillance programs previously established against these bacteria [14]. From this aspect, the Central Military Hospital in Ružomberok (CMH), Slovak Republic, the largest military medical facility in the territory of the Slovak Republic, which provides medical care not only to professional soldiers but also to the civilian population from the entire country, belongs to the category of important medical centres threatened by the arrival of COVID-19. We set ourselves the goal of retrospectively analysing the incidence of carbapenemase-producing *Klebsiella pneumoniae* in this hospital during the COVID-19 pandemic and to compare it with the pre-COVID-19 era. Consequently, the probable factors leading to changed CRKP incidence during the COVID-19 pandemic are discussed. The antimicrobial susceptibility of the CRKP strains from the COVID-19 era is evaluated to detect the frames of antimicrobial therapy choices. Based on the results, the antibiotic therapeutic options for CRKP-caused infections in CMH during the COVID-19 pandemic are estimated, and a complementary non-antibiotic solution employing phage therapy is considered. The CRKP isolates from 2022 were submitted to whole genome sequencing (WGS) to detect the genetic determinants of virulence and antimicrobial resistance and to identify the clonal affiliation, enabling a more exact evaluation of epidemiologic relations.

2. Results

In a retrospective study, the carbapenem-resistant *Klebsiella pneumoniae* (CRKP) incidence and distribution were analysed from 2017 to 2022, and the pre-COVID-19 period was compared with the COVID-19 era. Clinical samples of 28,808 patients, sent for microbiologic analysis, were included. The antimicrobial susceptibility of CRKP strains isolated during the COVID-19 pandemic was examined, and the strains isolated in 2022 were submitted to whole genome sequencing. All analyses excluded multiple isolations of the same CRKP strain from the same patient.

2.1. *The Incidence of Carbapenemase-Producing* Klebsiella pneumoniae

Clinical samples of 16,250 patients were submitted for microbiologic examination during the pre-COVID-19 period from 2017–2019. In 30 patients (0.18%), CRKP strains were isolated. In total, 3 strains produced KPC and 27 produced NDM type of carbapenemase. During the COVID-19 period (2020–2022), 12 558 patients were examined. Of them, 95 (0.76%) were CRKP-positive (for an overview, see Supplementary Tables S1 and S2). At the same time, 45 patients (47%) also had COVID-19. In total, 26 patients had KPC and 69 had NDM-carbapenemase-producing CRKP strains. The gender distribution was approximately equal, both in the basic and the CRKP-positive groups of patients (55% vs. 56.7% of females in 2017–2019; 52.6% vs. 49.5% in 2020–2022). A significant, 4.8-fold increase in CRKP relative incidence was detected during the COVID-19 period ($p < 0.0001$). NDM-producing CRKPs had a peak of their incidence in 2021, followed by a steep decrease. KPC-producing CRKPs have been slightly increasing since 2020 (Figure 1).

2.2. *Carbapenemase-Producing* Klebsiella pneumoniae *Distribution in Different Age Groups, Hospital Units, and Clinical Samples*

The age of CRKP-positive patients during the COVID-19 pandemic ranged from 2 to 91 years, with a median age of 67 years (more details in Supplementary Table S1). Patients older than 50 represented the majority of CRKP-positive cases in both examined time periods. In 2017–2019, they represented 90% of all CRKP-positive cases; in 2020–2022, this proportion was 83% (Figure 2).

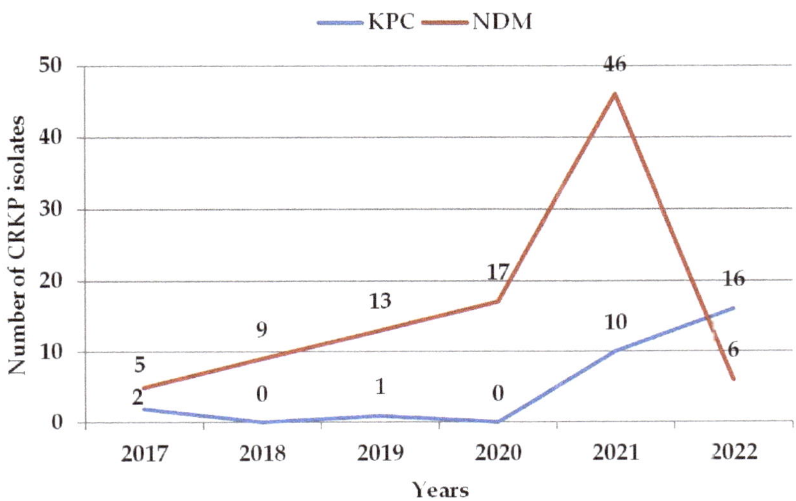

Figure 1. Incidence of carbapenemase-producing *Klebsiella pneumoniae* during the analysed years. CRKP—carbapenem-resistant *Klebsiella pneumoniae*; KPC—*Klebsiella pneumoniae* carbapenemase; NDM—New Delhi metallo-beta-lactamase. Note: multiple CRKP isolates from the same patient were excluded.

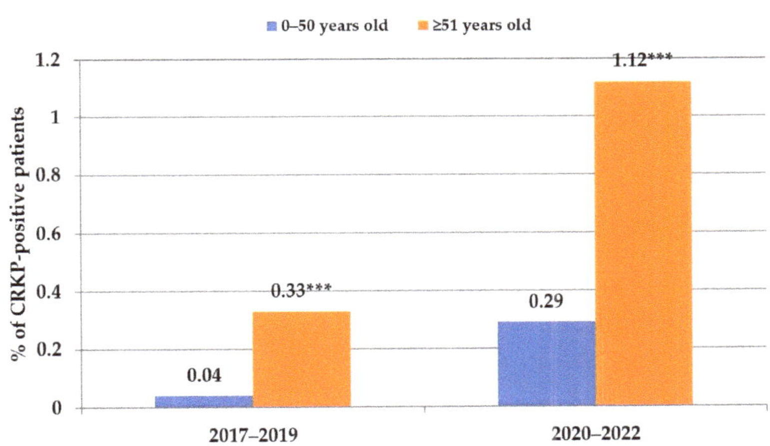

Figure 2. Age distribution of CRKP-positive patients with respect to risk-age for severe COVID. 0–50 years old: $p = 0.0003$; ≥51 years old: *** $p < 0.0001$.

During the analysed time periods, the proportion of CRKPs was excessively higher in the units with intensive-care patients (Figure 3). During the COVID-19 period, in 167 patients in intensive care units, *K. pneumoniae* was isolated from the lower respiratory tract as the cause of pneumonia. In total, 58 of them had CRKP strains (38 patients were COVID-19-positive), and 48 of them suffered from CRKP-caused ventilator-associated pneumonia (VAP) (34 were COVID-19-positive). In six patients with CRKP-caused VAP, invasion of *K. pneumoniae* into the bloodstream was detected; all patients suffered from COVID. A total of 31 patients with VAP died during hospitalisation, representing nearly 65% in-hospital mortality; 24 of them were COVID-19 patients (71% mortality rate) (Supplementary Table S2). In accordance with these data, the most notable increase in CRKP-positivity during the COVID-19 pandemic was detected in samples from the lower respiratory tract (from 0 to 2.6%). They were followed by samples from skin and soft-tissue infections (from 0.08% to 1.67%) and haemocultures (from 0.05 to 0.61%) (Figure 4).

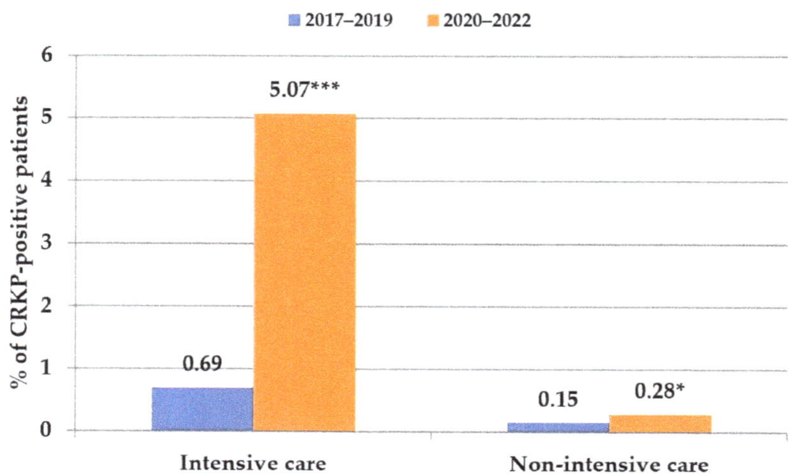

Figure 3. Carbapenem-resistant *Klebsiella pneumoniae* distribution in hospital units with intensive-care patients and in non-intensive-care units. * $p < 0.05$; *** $p < 0.0001$.

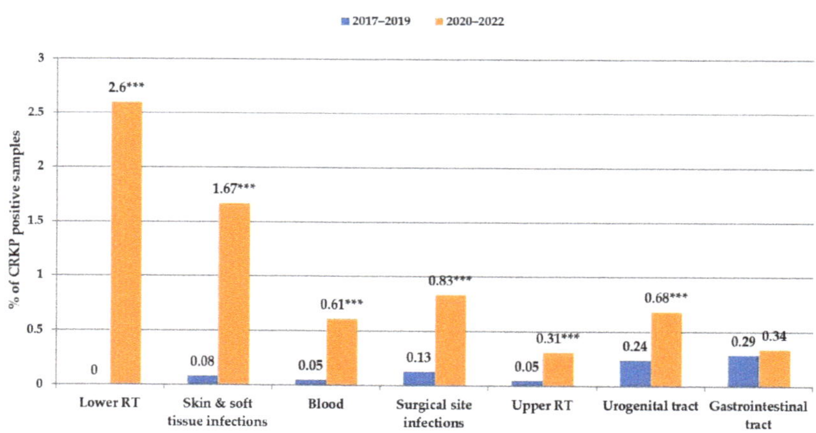

Figure 4. Carbapenem-resistant *Klebsiella pneumoniae* distribution in clinical samples from different body localities. *** $p < 0.0001$.

2.3. Antimicrobial Susceptibility of Carbapenemase-Producing Klebsiella pneumoniae

Antibiotic susceptibility results were analysed in the CRKPs from the COVID-19 period. Of the 18 tested antibiotics, 8 (ampicillin—intrinsic resistance; ampicillin–sulbactam, piperacillin–tazobactam, cefuroxime, cefotaxime, ceftazidime, cefepime, and ertapenem) were ineffective. The antimicrobial susceptibility to the rest of the antibiotics considerably differed between the KPC and NDM producers (Figures 5 and 6). Among the KPC-positive strains, the lowest rates of drug resistance were found for colistin (8%) and amikacin (15%), followed by gentamicin and tigecycline (both 88%). The lowest drug resistance in the NDM producers was again detected for colistin, but it was much higher than in KPC-strains: 43%. Colistin was followed by tigecycline, tetracycline, and co-trimoxazole (with 45, 48, and 71% resistance, respectively).

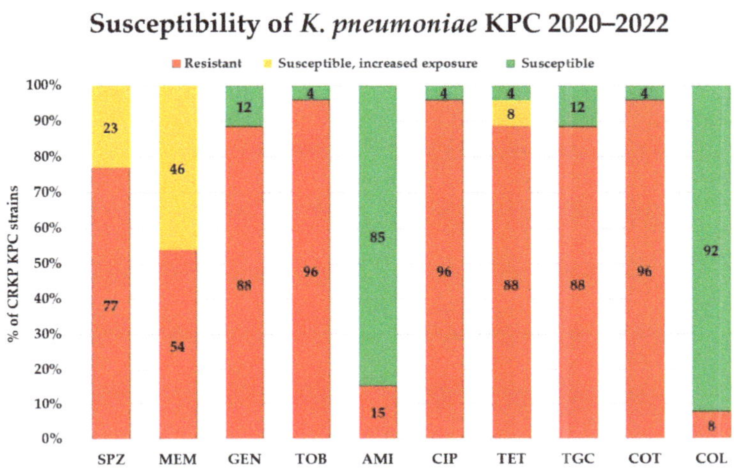

Figure 5. Susceptibility of KPC-producing *Klebsiella pneumoniae* isolated during COVID-19 pandemic. SPZ cefoperazone–sulbactam, MEM meropenem, GEN gentamicin, TOB tobramycin, AMI amikacin, CIP ciprofloxacin, TET tetracycline, TGC tigecycline, COT co-trimoxazole (trimethoprim–sulfamethoxazole), COL colistin. All isolates were resistant to ampicillin (intrinsic resistance); ampicillin–sulbactam, piperacillin–tazobactam, cefuroxime, cefotaxime, ceftazidime, cefepime, and ertapenem.

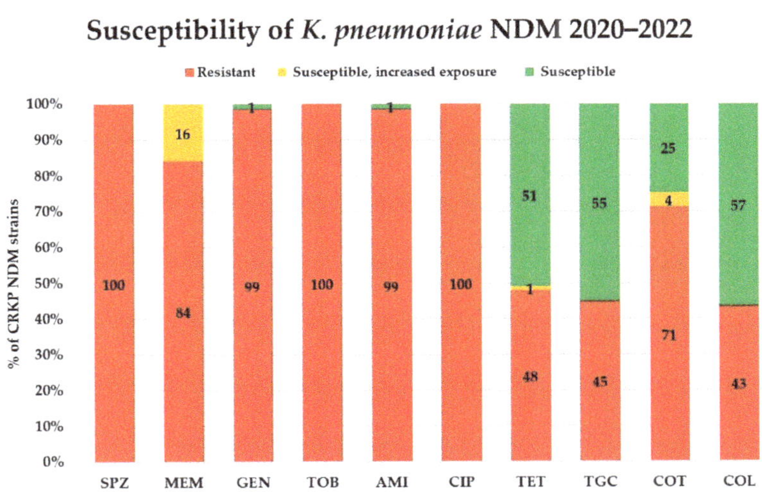

Figure 6. Susceptibility of NDM-producing *Klebsiella pneumoniae* isolated during the COVID-19 pandemic. SPZ cefoperazone–sulbactam, MEM meropenem, GEN gentamicin, TOB tobramycin, AMI amikacin, CIP ciprofloxacin, TET tetracycline, TGC tigecycline, COT co-trimoxazole (trimethoprim–sulfamethoxazole), COL colistin. All isolates were resistant to ampicillin (intrinsic resistance); ampicillin–sulbactam, piperacillin–tazobactam, cefuroxime, cefotaxime, ceftazidime, cefepime, and ertapenem.

The distribution of MIC values for meropenem (breakpoint 2 mg/L for category "susceptible", 8 mg/L for category "susceptible, increased exposure"), amikacin (breakpoint 8 mg/L) and colistin (breakpoint 0.2 mg/L) are shown in Supplementary Figures S1–S3. Meropenem had MICs for a comparatively large subset of KPC strains from the interval 2–8 mg/L (susceptible, increased exposure); similarly, the distribution of MIC values of colistin and amikacin in most KPC strains was closer to the breakpoint than among the NDM strains.

2.4. *In Vitro Activity of the New Beta-Lactams*

The activity of one new cephalosporin and three new combinations of beta-lactams with inhibitors of beta-lactamases was tested in the group of 12 CRKP strains isolated in the year 2022. KPC-producing isolates were tested against cefiderocol (breakpoint 2 mg/L), ceftazidime–avibactam (breakpoint 8 mg/L), imipenem–relebactam (breakpoint 2 mg/L) and meropenem–vaborbaktam (breakpoint 8 mg/L). All four therapeutics were designed to treat infections caused by multiresistant Gram-negative bacteria, including CRKPs. The NDM-producing isolates were tested only against cefiderocol, as the inhibitors of beta-lactamase combined with the other three beta-lactam drugs are inactive against class B metallo-beta-lactamases. An excellent in vitro activity was detected in all analysed isolates. Even if cefiderocol MIC values were slightly higher in the NDM-producing strains, they were considerably far from the breakpoints (Tables 1 and 2).

Table 1. Ceftazidime–avibactam, imipenem–relebactam and meropenem–vaborbaktam in vitro activity against KPC-producing carbapenem-resistant *Klebsiella pneumoniae*.

Ceftazidime–Avibactam MIC [mg/L]	KPC (n)	Imipenem–Relebactam MIC [mg/L]	KPC (n)	Meropenem–Vaborbaktam MIC [mg/L]	KPC (n)
0.032	1	0.094	1	0.023	3
0.094	1	0.125	2	0.032	3
0.190	2	0.190	1	0.047	1
0.380	1	0.250	1		
0.750	2	0.380	1		
		0.5	1		

MIC—minimal inhibitory concentration; KPC—*Klebsiella pneumoniae* carbapenemase.

Table 2. Cefiderocol in vitro activity against carbapenem-resistant *Klebsiella pneumoniae*.

Cefiderocol MIC [mg/L]	KPC (n)	NDM (n)
0.016	1	-
0.032	4	-
0.064	1	1
0.094	1	1
0.19	-	2
0.25	-	1

MIC—minimal inhibitory concentration; KPC—*Klebsiella pneumoniae* carbapenemase; NDM—New Delhi metallo-beta-lactamase.

2.5. *Genome Sequencing of Selected Carbapenem-Resistant* Klebsiella pneumoniae *Strains*

Twelve CRKP isolates from 2022 were analysed via WGS. The length of the genomes ranged from 5.38 to 6.04 Mbp, and when analysed with MLST, the isolates belonged to three sequence types (STs) (Figure 7, Table 3). Five strains belonged to ST-11 of sublineage SL-258, the other six were ST-307 and SL-307, and the last strain was of the rare type ST-5889. The mutual relatedness of strains assigned to the same ST was further assessed using core genome MLST (cgMLST). Strains were clustered according to the sequence types (Figure 7).

Four of the five ST-11 strains were mutually closely related as they differed by less than four alleles; the last strain was different as it possessed 95 distinct alleles and belonged to different clonal group (Supplementary Table S3). Strains of ST-307 were very homogenous, and they had almost no difference between each other.

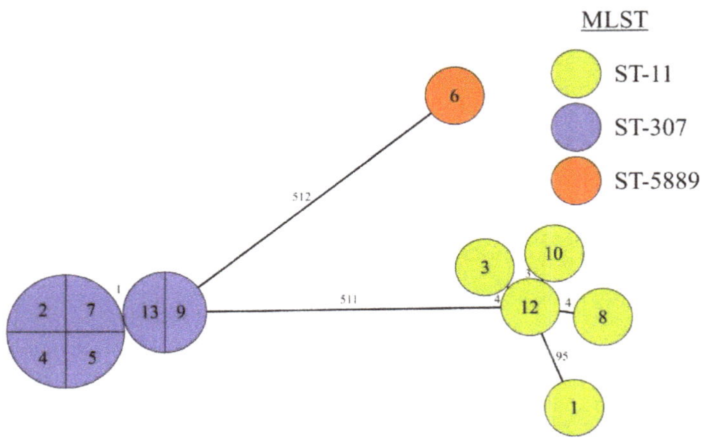

Figure 7. GrapeTree of 12 carbapenem-resistant *Klebsiella pneumoniae* isolates based on cgMLST. Ring sizes correspond to the number of strains with the same genotype; the designation of the strains is marked inside the ring, and the ST are designated by colour. The numbers on the lines between rings correspond to the numbers of different alleles.

Table 3. Characterisation of 12 recent isolates of carbapenem-resistant *Klebsiella pneumoniae* with sequenced genomes.

Sublineage	ST	Clonal Group	Serotype (wzi) [1]	Carbapenemase Gene	No. of Strains
SL-258	11	CG340	K15:O4 (50)	NDM-1	4
		CG3666	? (24)	NDM-1	1
SL-307	307	CG307	K2:O1/2 (173)	KPC-2	6
SL-10716	5889	CG11340	K14/15:O3 (new)	KPC-2	1

[1] Serotype determined according to K-PAM (*wzi* gene allele number).

The presence of antibiotic resistance genes was analysed using the BV-BRC server by using the NDARO database. Five to eighteen resistance genes were detected in individual strains (Supplementary Table S4). All ST11 strains produced metallo-beta-lactamase NDM-1 and the serine beta-lactamase SHV-11; three of them also contained genes for CTX-M-15 and OXA-1. The most different ST-11 strain in cgMLST also differed in the antibiotic resistance profile. Strains belonging to ST-307 were very similar in their antibiotic resistance gene profiles. Besides KPC-2 beta-lactamase, they harboured genes for CTX-M-15, SHV-28, TEM-1, and OXA-1 beta-lactamases. The unique ST-5889 strain had the fewest resistance genes. It contained KPC-2 and SHV-1 beta-lactamases genes. The antimicrobial resistance phenotype pattern and the corresponding sequence types of the analysed carbapenemase-producing *Klebsiella pneumoniae* are shown in Figure 8.

The virulence factors in the sequenced strains were monitored using the *Klebsiella* BIGSdb database. Relatively low virulence potential was observed (Supplementary Table S5). All strains contained *mrk* operon encoding for Type 3 fimbriae and *iut*A and *glx*K genes. In addition, ST-11 strains possessed yersiniabactin genes.

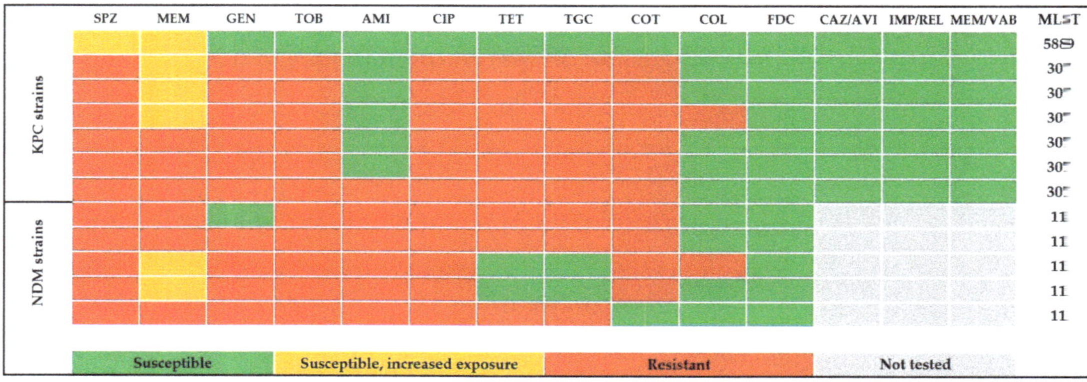

Figure 8. Antibiogram and the sequence-types distribution in carbapenem-resistant *Klebsiella pneumoniae* from the year 2022. SPZ cefoperazone–sulbactam, MEM meropenem, GEN gentamicin, TOB tobramycin, AMI amikacin, CIP ciprofloxacin, TET tetracycline, TGC tigecycline, COT co-trimoxazole (trimethoprim–sulfamethoxazole), COL colistin, FDC cefiderocol, CAZ/AVI ceftazidime–avibactam, IMP/REL imipenem–relebactam, MEM/VAB meropenem–vaborbaktam, KPC *Klebsiella pneumoniae* carbapenemase, NDM—New Delhi metallo-beta-lactamase. All isolates were resistant to ampicillin (intrinsic resistance); ampicillin–sulbactam, piperacillin–tazobactam, cefuroxime, cefotaxime, ceftazidime, cefepime, and ertapenem.

3. Discussion

The survey of nosocomial strains and analysis of their incidence and antimicrobial susceptibility is vital for the treatment and prevention of hospital-acquired infections. Of them, infections caused by CRKPs are of enormous importance. During the COVID-19 pandemic, increased numbers of CRKP-caused nosocomial bacterial superinfections were reported from various countries worldwide [10–13,15]. Our Central Military Hospital (CMH) was no exception: the carbapenem-resistant *Klebsiella pneumoniae* relative incidence significantly increased by 4.8 times. However, our results are in contradiction with some Chinese and Italian studies, which have not detected any statistically significant difference in the incidence ratio, colonisation rate, and infection rate of carbapenem-resistant Enterobacterales during the pandemic [16,17]. Almost half of the CRKP-positive patients of the CMH had a COVID-19 infection at the same time.

People over 50 belong to the risk group for severe COVID-19 infection [18], and COVID-19 patients are especially threatened by secondary CRKP infection due to their extreme fragility and need for prolonged ICU stay [19]. Therefore, we tried to look at the age distribution of CRKP-positive patients. During the COVID-19 period (2020–2022), this age group represented 79% of all CRKP-positive patients. Even if CRKPs were isolated from patients over 50 with a significantly higher frequency in the COVID-19 period, similar results were obtained in the pre-COVID-19 era. In this respect, older patients may have several other individual risk factors contributing to infection caused by CRKP [20], as also may have some young patients with severe COVID-19 [19]. In our study, there were no significant differences in the incidence of CRKPs among male and female patients, in contrast with Yang et al., who detected higher CRKP proportions and resistance in male patients [16]. Similar results were presented by Mędrzycka-Dąbrowska et al. in their extensive review of CRKP infections in COVID-patients [21], who reported that most CRKP-patients were males with a mean age of 61 years.

The incidence of CRKP isolates was significantly higher in the units with intensive-care patients, in concordance with reports of many other studies; for review, see [21]. This phenomenon was even more pronounced during the COVID-19 period when conditions in the CMH changed. There were higher numbers of seriously ill patients, mainly with pneumonia, often with the need for artificial lung ventilation (ALV), which increased the risk of secondary bacterial pneumonia. In CMH, a nearly 71% mortality rate was registered among the patients with COVID-19 on ALV with CRKP-caused secondary bacterial pneumonia. Similar mortality rates were reported in a retrospective, single-centre case–control study from Spain, where carbapenemase-producing Enterobacterales were the cause of death in about two-thirds of the critically ill patients with COVID-19 [22]. The study by Dumitru et al. reports nine ICU patients with severe invasive CRKP co-infection with COVID-19. All patients but one were over 50, with considerable comorbidities. Five patients died, representing over 50% mortality rate [23].

Parallel with the increase in CMH patients on ALV, the CRKP-positivity of the sputum samples also increased; likewise, the Chinese study identified sputum as the primary source of CRKPs during the COVID-19 pandemic [16]. The major reservoir of *K. pneumoniae* infections is the gastrointestinal carriage [24]; however, the incidence of CRKP in gastrointestinal samples of CMH patients during the COVID-19 period increased only slightly. Artificial lung ventilation represents an increased risk of secondary nosocomial lung infection, which can be further worsened by primary viral pulmonary infection. Even if CRKP finding in the upper respiratory tract is more indicative of colonisation, it can become a source of lung invasion during intubation [25]. Nevertheless, our study revealed only a moderate increase in the incidence of CRKP in the samples from the upper respiratory tract. Several factors were revealed through analysing the potential reasons for a steep increase in the incidence of CRKP in the CMH during 2021, which primarily affected the patients of the Department of Anaesthesiology and Intensive Care Medicine (DAICM). In 2021, the number of patients with artificial lung ventilation increased four times compared to the pre-COVID-19 period. It resulted in a huge burden for the staff and a shortage of nurses and doctors. A feeling of false cleanliness could also play an important role. The staff were dressed in overalls with two gloves. Still, MDR *Klebsiella* strains can survive disinfection and remain on the overalls, allowing the spread from patient to patient, whereas there were patients from all over the Slovak Republic hospitalised in the CMH DAICM. In addition, the DAICM moved several times, for technical and organisational reasons, to different hospital areas. It is known that in such unprecedented conditions, the inability to conform completely to standard infection control practices may contribute to a shift in antimicrobial resistance, including resistance to carbapenems [17], as confirmed by the analysis of carbapenem-resistant Enterobacterales (CRE) incidence in the Italian study by Tiri et al. [15]. They detected a CRE acquisition increase from 6.7% in 2019 to 50% in March–April 2020 (the most critical COVID-19 period in Italy) and identified several factors that contributed to CRE spreading: the high intensity of care, the prone position requiring 4–5 healthcare workers equipped with personal protective equipment in a high-risk area and with extended and prolonged contact with the patient, and the presence of healthcare workers from other departments without work experience in the ICU setting. These findings were in many aspects similar to ours. Over the next year, the strain of patients in the CMH reduced, and their number decreased to half that from 2021. Moreover, the DAICM staff learned from the previous crisis. Stricter protection measures have been introduced, including wearing a disposable gown on top of the overalls during care for patients colonised by MDR strains, increased hand hygiene and extra gloves, isolation or cohortation of CRKP-positive patients (as the patient burden and the hospital capacities allowed), re-education of the staff, environmental surface monitoring for CRKP, immediate reporting of new positive cases, and increased hospital hygiene and asepsis. This resulted in a drop in NDM-producing CRKP incidence to the levels of the pre-pandemic period; however, KPC-producing CRKP appeared and started to increase slightly but steadily, even if it still did not indicate a situation of nosocomial outbreak.

CRKPs from the year 2022 were submitted to whole genome sequencing. Based on sequencing results, carbapenem-resistant *K. pneumoniae* isolated in the CMH represented a phylogenetically not very diverse group, which suggests an epidemic nosocomial spread. The majority of the isolates were of ST-11 (NDM-1 producers) and ST-307 (KPC-2 producers). Only one KPC-2 isolate was of a different ST type: ST-5889. In Europe, CRKP STs 11, 15,

101, and 258/512 are widely distributed [1]. Moreover, ST-11 is the dominant CRKP clone in China, with an alarming spread [26], successful international transmission [27], and capacity to be converted to hypervirulent (hv) carbapenemase-producing infectious agents, as demonstrated by the recently emerged ST-11 CR-HvKP strains [28]. On the other hand, CRKPs of ST-11 may also lose antimicrobial resistance and virulence [29], which means that their genome must be very plastic and requires permanent monitoring; however, this feature is typical for many other *Klebsiella pneumoniae* sequence types. *Klebsiella pneumoniae* ST-307 emerged in 2008 and, since that time, has been disseminated to different parts of the world, causing hospital outbreaks [30]. There is a lack of literary data on a unique CRKP ST-5889. Only one more isolate (from China) is deposited in the *Klebsiella* BIGSdb database. Our isolate of CRKP ST-5889 substantially differed from the other CRKPs from the CMH in a lower number of beta-lactamases and a higher susceptibility to non-beta-lactam drugs. Only one study is available concerning CRKP sequence type distribution in other hospitals in the Slovak Republic. A little broader spectrum of CRKP sequence types was present in the three medical facilities of the University Hospital in Bratislava, Slovak Republic, as reported by Koreň et al. [4]. Similar to our results, the majority of CRKPs were ST-11 NDM-1 producers. They were followed by KPC-2 producers of ST-258—a dominant ST in CRKP in Europe and America [1]—and infrequent ST-584. Just one isolate was reported from the ST-15 and ST-340, and none of our hospital's most abundant KPC-2-producing ST-307. Our strains did not contain any genes for the production of hypermucoid capsule or siderophores other than yersiniabactin, which are important factors of hypervirulent *Klebsiella* clones [28]. This observation is in agreement with a nosocomial spread of the strains.

Treatment of patients infected with CRKP represents a big challenge. Depending on the type of produced carbapenemase, these strains are resistant to various beta-lactam drugs, and their susceptibility to non-beta-lactam drugs is not predictable due to the acquisition of different additional resistance mechanisms [31,32], as our results also documented it. Therefore, the therapeutical options are limited, and the treatment response is uncertain [33]. *Klebsiella pneumoniae* carbapenemase (KPC) and metallo-beta-lactamases, such as NDM, represent the basic molecular mechanism of CRKP resistance to carbapenems [34]. CRKPs in our study produced only these two classes of carbapenemases. Different antimicrobial resistance patterns were identified among KPC and NDM producers. The most considerable difference was in the susceptibility to amikacin, with mostly susceptible KPC producers and resistance in almost all NDM producers. Substantial differences were also detected in susceptibility to colistin—an "old" drug currently experiencing a renaissance in clinical use, driven by the increased spreading of multidrug-resistant Gram-negative rods in hospital settings. Even if rather toxic, colistin was considered, in many cases, a last-resort drug. Due to the spread of antimicrobial resistance and the appearance of CRKP, it is now often used as a first-line treatment for infections caused by these bacteria [35]. With more and more frequent use of colistin in therapy, colistin-resistant strains with modified LPS molecules have started to emerge [36]. In our study, the KPC-producing CRKP strains had only a low colistin-resistance rate; however, we did not expect such high colistin resistance in the NDM producers (over 40%). Also, the colistin-MIC distribution was less favourable for NDM-strains, with values far from the breakpoint, which may be a disadvantage when a combination treatment is considered. Growing numbers of colistin-resistant CRKP isolates were also reported from University Hospital facilities in the Slovak Republic, even if not so alarming—from 0% published in 2019 to 2.5% in 2022 [3,4]. When analysing this extremely high colistin-resistance rate in the CMH among the NDM CRKP strains, the most probable reason is the epidemic spread of the same colistin-resistant NDM *Klebsiella pneumoniae* strain during the COVID-19 period (as discussed above). Based on in vitro studies [37] and clinical trials [38], colistin in combination with either meropenem or amikacin could be a valid therapeutic option against colistin-resistant CRKP isolates. Unfortunately, this option could not have been applied in most CMH patients with NDM-producing CRKPs due to their high resistance rate against these antimicrobial drugs. Tigecycline is another drug with activity against CRKP and a positive effect on the survival of patients with severe CRKP-caused infections when applied in high doses in combination therapy [39]. However, resistance emerging has been described during the treatment of CRKP infections [40]. In our study, based on the susceptibility test results, the successful effect of the combination therapy with tigecycline could be expected in more than half of the infections caused by NDM-producing CRKPs, but only in less than 20% of cases caused by KPC-producers. The unusual combination of ertapenem with another carbapenem in the therapy of CRKP-caused infections is based on the evidence that ertapenem could bind to the carbapenemase in a very high affinity which prevents the hydrolysis of the other carbapenem molecule and preserves its bactericidal activity [41]. From meropenem's MIC distributions, we guess this therapeutical option could work better for KPC producers, which had MICs distribution closer to the breakpoint.

Correct identification of carbapenemase type is crucial for selecting the most suitable antimicrobial agent among those recently introduced to the therapeutic practice [34]. In general, CRKPs, which produce metallo-beta-lactamase (including NDM), are resistant to all beta-lactams (and their combinations with beta-lactamase inhibitors), except cefiderocol—a first siderophore cephalosporin—and the monobactam aztreonam [42]. In metallo-beta-lactamase-producing CRKP, aztreonam can frequently be inactivated by other beta-lactamases. A combination of ceftazidime–avibactam and aztreonam can have a synergistic effect, as avibactam can inactivate other beta-lactamases to preserve aztreonam activity [43,44]. However, testing of these two antimicrobials was not included in our CRKP strains' routine susceptibility testing scheme.

For therapy of infections caused by KPC producers, combinations of beta-lactams with new beta-lactamase inhibitors, such as ceftazidime–avibactam, meropenem–vaborbactam, or imipenem–relebactam, or a new cefalosporin, cefiderocol, are suggested [45]. These new drugs are still only seldom used in the Slovak Republic. Their rare use is in accordance with the excellent susceptibility detected in CRKPs isolated in 2022 in CMH. Unfortunately, there are already reports on resistance emergence against these new valuable drugs [45,46]. Clinical failure of treatment with ceftazidime–avibactam was described by Dumitru et al. [25] in patients whose diagnosis was late. Solutions may be the early and correct identification of the CRKP-produced carbapenemase type, early therapy with effective drugs, strict adherence to rational antimicrobial therapy, and a judicious prescription of these valuable antimicrobials. These precautions may slow down the emergence and spread of resistant strains and preserve the activity of the drugs mentioned above for the future [45,46]. Alternative or complementary therapies—such as phage therapy—can also be helpful in these efforts. Several reports of COVID-19 patients' treatment with therapeutic phages were published during the COVID-19 pandemic. COVID-19 patients with secondary bacterial infections were successfully treated with bacteriophages in China [47] and Russia [48], and the FDA has also allowed the use of phage therapy for COVID-19 patients who are in a critical state [49]. This old and nowadays rediscovered antibacterial therapy has many advantages over traditional antibiotics. Phages are highly specific to their hosts, preserving the microbiota equilibrium during treatment. In the infectious focus, they amplify and may spread and attack their specific hosts over the body without respect to antimicrobial resistance mechanisms and do not select for antimicrobial resistance [50]. Phage therapy is usually safe, but the phage-induced release of endotoxins may lead to an immune reaction, depending on the phages, delivery routes, or host immune status, as it was observed by Wu et al. (2022). One of their COVID-19 patients developed transient fever and IL-6 and IL-8 storm four hours after phage inhalation; such reaction might be a result of triggering human immune responses directly by antigens of therapeutic phages or indirectly via the rapid lysis of the targeted bacteria [47].

4. Materials and Methods

4.1. Data Resources

The laboratory information system database of the Institute of Clinical Microbiology, Central Military Hospital in Ružomberok, Slovak Republic, was analysed for data on

Klebsiella pneumoniae isolations during 2017–2022. The incidence of CRKP strains in the COVID-19 and pre-COVID-19 periods (years 2020–2022 and 2017–2019, respectively) and the antimicrobial susceptibility of CRKP strains during the COVID-19 period were then retrospectively analysed. Data on CRKP strains distribution among patients, hospital units, and biological sample types were also examined. Multiple isolations of the same strain from the same patient were excluded from the analysis. The patients' personal data protection, in conformity with the GDPR, was applied during the analysis.

4.2. *Bacterial Strains Isolation, Identification and Long-Term Preservation*

The clinical samples were collected, transported, and processed in conformity with the standard diagnostic procedures. The identification of *Klebsiella pneumoniae* strains was carried out using the biochemical ENTEROtest 16 kit (Erba Lachema s.r.o., Brno, Czech Republic). Ambiguous results were confirmed using MALDI-TOF MS. All CRKPs isolated during 2022 were long-term preserved by freezing in cryoprotective medium (5% *w*/*v* Skim Milk, Biolife, Milan, Italy) at −20 °C.

4.3. *Antimicrobial Susceptibility Testing and Carbapenemase Type Detection*

A colourimetric micro-method for automated susceptibility testing MIDITECH (Bell Novaman, Bratislava, Slovak Republic), meeting the antimicrobial susceptibility testing recommendations of the CLSI [51], was used for susceptibility testing to 18 antimicrobial agents (ampicillin, ampicillin–sulbactam, piperacillin–tazobactam, cefuroxime, cefotaxime, ceftazidime, cefoperazone–sulbactam, cefepime, ertapenem, meropenem, gentamicin, tobramycin, amikacin, tigecycline, ciprofloxacin, tetracycline, colistin, and co-trimoxazole) and 2 cephalosporin combinations with inhibitors of beta-lactamase for detection of extended-spectrum beta-lactamase production (cefotaxime-clavulanate, ceftazidime-clavulanate). E-test (Liofilchem, S.r.l., Roseto degli Abruzzi, Italy) was used to detect the susceptibility to cefiderocol, imipenem–relebactam, ceftazidime–avibactam, and meropenem–vaborbactam. MIC values were interpreted according to the EUCAST clinical breakpoints [52]. Screening of carbapenemase-positive strains and detection of carbapenemase production (using chromogenic pH metric Carba NP test; Diagnostics, s.r.o., Galanta, Slovak Republic) was performed according to EUCAST guidelines [53]. The rapid phenotypic identification of the carbapenemase type was made via immunochromatographic CARBA-5 assay (NG-Biotech Laboratoires, Guipry-Messac, France), designed for the detection of five major carbapenemases types (KPC, NDM, VIM, IMP, and OXA-48). Respecting the guidelines of the Ministry of Health of the Slovak Republic, the carbapenemase-producing *Klebsiella pneumoniae* strains were sent to the National Reference Centre for Antimicrobial Resistance Surveillance for confirmation.

4.4. *Genome Analysing of Klebsiella pneumoniae Isolates*

Overnight bacterial cultures grown on Luria–Bertani (LB) medium at 37 °C were used. DNA was isolated using the Higher PurityTM Bacterial Genomic DNA Isolation Kit (CanvaxBiotech; Cordoba, Spain) and was quantified using the Qubit ds-DNA HS Assay Kit (Thermofisher Scientific; Eugene, OR, USA). Sequencing libraries were preprepared using the Nextera XT DNA Library Prep Kit protocol (Illumina; San Diego, CA, USA) and purified on AMPure XP magnetic beads (Beckman Coulter Life Sciences; Indianapolis, IN, USA). The quality of the library was checked using a high-sensitivity DNA electrophoresis chip and a 2100 Bioanalyzer Instrument (Agilent Technologies; Santa Clara, CA, USA). Sequencing was carried out with 2 × 150 bp reads using the Illumina NextSeq 500 platform (Illumina). The obtained sequencing data were assembled de novo with SPAdes using a standard setting of parameters (Center for Algorithmic Biotechnology; St. Petersburg State University, St Petersburg, Russia) [54] and annotated using the online BV-BRC program (Bacterial and Viral Bioinformatics Resource Center, https://www.bv-brc.org (accessed on 14 July 2023). The criterion of 100% identity and 100% coverage with the reference gene was used as the threshold for gene presence. The genomes were further analysed using the CGE database (Center for Genomic Epidemiology; Kongens, Lyngby, Denmark), the BV-BRC server, and the BIGSdb database (Institut Pasteur MLST, Paris, France). The standard MLST and cgMLST based on 629 genes available in the *Klebsiella* BIGSdb database (https://bigsdb.pasteur.fr/klebsiella/, (accessed on 16 July 2023) were used for the determination of strain relatedness. GrapeTree was used to visualise strain clusters based on the *Klebsiella* cgMLST analysis. The sequenced genomes were deposited in the NCBI database under bioproject accession number PRJNA996751 and into the BIGSdb-Pasteur *Klebsiella pneumoniae* database under accession numbers 57223–57235.

4.5. Statistical Analysis

The results were statistically analysed with the chi-square test using Excel software version 2019 (Microsoft Corporation, Redmond, WA, USA). p values of <0.05 were considered statistically significant.

5. Conclusions

Isolates of CRKP increased in the CMH during the COVID-19 pandemic. The molecular mechanisms responsible for carbapenem resistance were KPC and NDM carbapenemase production. The recent isolates had relatively low phylogenetic diversity, which may indicate an epidemical spread; all were susceptible to new antimicrobials appointed to treat CRKP-caused infections. Despite the excellent activity of new antimicrobials, a rational antibiotic policy must be thoroughly followed, supported by complementary treatments and strict anti-epidemic precautions. Based on the available literary data, bacteriophages, with their high specificity and rare side effects, would be of great value in the therapy of nosocomial bacterial superinfections caused by CRKP strains, and not only in COVID-19 patients.

Supplementary Materials: The following supporting information can be downloaded at: https://www.mdpi.com/article/10.3390/antibiotics12081285/s1, Figure S1: Meropenem MIC values distribution in KPC (a) and NDM (b) producing carbapenemase-resistant *Klebsiella pneumoniae;* Figure S2: Amikacin MIC values distribution in KPC (a) and NDM (b) producing carbapenemase-resistant *Klebsiella pneumoniae;* Figure S3: Colistin MIC values distribution in KPC (a) and NDM (b) producing carbapenemase-resistant *Klebsiella pneumoniae;* Table S1: Distribution of patients with carbapenem-resistant *Klebsiella pneumoniae* during COVID-19 pandemic to age categories; Table S2: Patients infected by carbapenem-resistant *Klebsiella pneumoniae* during the COVID-19 pandemic; Table S3: Taxonomic characterization of sequenced *Klebsiella pneumoniae* strains; Table S4: Genes encoding antibiotic resistance in sequenced *Klebsiella pneumoniae* strains; Table S5: Virulence factor genes in sequenced *Klebsiella pneumoniae* strains.

Author Contributions: Conceptualisation, L.S. and H.D.; methodology, J.F., L.S. and H.D.; software, J.F., M.A., M.B. and T.S.; validation, J.F.; formal analysis, J.F. and M.A.; investigation, J.F., M.A. and M.B.; data curation, J.F., M.A., M.B. and T.S.; writing—original draft preparation, J.F.; writing—review and editing, L.S., H.D. and A.L.; visualisation, J.F.; supervision, L.S., H.D. and T.S.; project administration, A.L. and L.S.; funding acquisition, L.S., A.L. and H.D. All authors have read and agreed to the published version of the manuscript.

Funding: This research was funded by the Operational Program Integrated Infrastructure for the project "Research and development in medical sciences—the way to personalise the treatment of serious neurological, cardiovascular, and cancer diseases" (code: ITMS: 313011T431), co-financed by the resources of the European Regional Development Fund; by the Slovak Research and Development Agency under the contract no. APVV-16-0168; and by the Slovak Republic Ministry of Education grant KEGA no. 002UK-4/2022.

Institutional Review Board Statement: The study was conducted in accordance with the Declaration of Helsinki, and approved by the Ethics Committee of Central Military Hospital in Ružomberok (ÚVN-56-41/2023, 26 June 2023).

Informed Consent Statement: Not applicable.

Data Availability Statement: The data for this study have been deposited in the NCBI database under bioproject accession number PRJNA996751 and into The BIGSdb-Pasteur *Klebsiella pneumoniae* database under accession numbers 57223–57235.

Acknowledgments: The authors wish to thank Katarína Ondrášiková and Zuzana Likavcová from the Institute of clinical microbiology, CMH in Ružomberok, Gizela Skýpalová from the Institute of Microbiology, Faculty of Medicine, Comenius University in Bratislava, and Tatiana Štrbová from the Department of Molecular Biology, Faculty of Natural Sciences, Comenius University in Bratislava, for routine laboratory diagnostic and technical cooperation, and Daniel Čema, from the Department of Anaesthesiology and Intensive Care Medicine CMH in Ružomberok, for clinical information and consultation on intensive-care CRKP-positive patients.

Conflicts of Interest: The authors declare no conflict of interest.

References

1. David, S.; Reuter, S.; Harris, S.R.; Glasner, C.; Feltwell, T.; Argimon, S.; Abudahab, K.; Goater, R.; Giani, T.; Errico, G.; et Epidemic of carbapenem-resistant *Klebsiella pneumoniae* in Europe is driven by nosocomial spread. *Nat. Microbiol.* **2019**, 1919–1929. [CrossRef] [PubMed]
2. Wang, N.; Zhan, M.; Liu, J.; Wang, Y.; Hou, Y.; Li, C.; Li, J.; Han, X.; Liu, J.; Chen, Y.; et al. Prevalence of Carbapenem-Resista *Klebsiella pneumoniae* Infection in a Northern Province in China: Clinical Characteristics, Drug Resistance, and Geograph Distribution. *Infect. Drug Resist.* **2022**, *15*, 569–579. [CrossRef] [PubMed]
3. Koreň, J.; Andrezál, M.; Drahovská, H.; Hubenáková, Z.; Liptáková, A.; Maliar, T. Next-Generation Sequencing of Carbapene Resistant *Klebsiella pneumoniae* Strains Isolated from Patients Hospitalized in the University Hospital Facilities. *Antibiotics* **202** *11*, 1538. [CrossRef] [PubMed]
4. Koren, J.; Hubenakova, Z.; Drahovska, H.; Ozaee, E.; Markuskova, B.; Lichvarikova, A. Emergence of extended-spectru beta-lactamase (ESBL) and/or carbapenemase producing *Enterobacteriaceae* (CPE) and their antimicrobial resistance. *Bratisl. L Listy* **2019**, *120*, 935–940.
5. Avendano, E.E.; Raman, G.; Chan, J.; McCann, E. Burden of carbapenem non-susceptible infections in high-risk patien Systematic literature review and meta-analysis. *Antimicrob. Resist. Infect. Control* **2020**, *9*, 193. [CrossRef]
6. Wang, F.; Zou, X.; Zhou, B.; Yin, T.; Wang, P. Clinical characteristics of carbapenem-resistant *Klebsiella pneumoniae* infe tion/colonisation in the intensive care unit: A 9-year retrospective study. *BMJ Open* **2023**, *13*, e065786. [CrossRef]
7. Zhen, X.; Stålsby Lundborg, C.; Sun, X.; Gu, S.; Dong, H. Clinical and Economic Burden of Carbapenem-Resistant Infection Colonization Caused by *Klebsiella pneumoniae, Pseudomonas aeruginosa, Acinetobacter baumannii*: A Multicenter Study in Chir *Antibiotics* **2020**, *9*, 514. [CrossRef]
8. Dandachi, I.; Azar, E.; Hamouch, R.; Maliha, P.; Abdallah, S.; Kanaan, E.; Badawi, R.; Khairallah, T.; Matar, G.M.; Daoud, *Acinetobacter* spp. in a Third World Country with Socio-economic and Immigrants Challenges. *J. Infect. Dev. Ctries* **2019**, 948–955. [CrossRef]
9. Arcari, G.; Raponi, G.; Sacco, F.; Bibbolino, G.; Di Lella, F.M.; Alessandri, F.; Coletti, M.; Trancassini, M.; Deales, A.; Puglies F.; et al. *Klebsiella pneumoniae* infections in COVID-19 patients: A 2-month retrospective analysis in an Italian hospital. *Int. Antimicrob. Agents* **2021**, *57*, 106245. [CrossRef]
10. Chatterjee, N.; Nirwan, P.K.; Srivastava, S.; Rati, R.; Sharma, L.; Sharma, P.; Dwivedi, P.; Jaggi, N. Trends in carbapenem resistan in Pre-COVID and COVID times in a tertiary care hospital in North India. *Ann. Clin. Microbiol. Antimicrob.* **2023**, *22*, 1. [CrossRe
11. Despotovic, A.; Milosevic, B.; Cirkovic, A.; Vujovic, A.; Cucanic, K.; Cucanic, T.; Stevanovic, G. The Impact of COVID-19 on th Profile of Hospital-Acquired Infections in Adult Intensive Care Units. *Antibiotics* **2021**, *10*, 1146. [CrossRef] [PubMed]
12. Karataş, M.; Yaşar-Duman, M.; Tünger, A.; Çilli, F.; Aydemir, Ş.; Özenci, V. Secondary bacterial infections and antimicrobi resistance in COVID-19: Comparative evaluation of pre-pandemic and pandemic-era, a retrospective single center study. *An Clin. Microbiol. Antimicrob.* **2021**, *20*, 51. [CrossRef] [PubMed]
13. Karruli, A.; Boccia, F.; Gagliardi, M.; Patauner, F.; Ursi, M.P.; Sommese, P.; De Rosa, R.; Murino, P.; Ruocco, G.; Corcione, A.; et Multidrug-Resistant Infections and Outcome of Critically Ill Patients with Coronavirus Disease 2019: A Single Center Experien *Microb. Drug Resist.* **2021**, *27*, 1167–1175. [CrossRef] [PubMed]
14. Belvisi, V.; Del Borgo, C.; Vita, S.; Redaelli, P.; Dolce, P.; Pacella, D.; Kertusha, B.; Carraro, A.; Marocco, R.; De Masi, M et al. Impact of SARS CoV-2 pandemic on carbapenemase-producing *Klebsiella pneumoniae* prevention and control programm Convergent or divergent action? *J. Hosp. Infect.* **2021**, *109*, 29–31. [CrossRef]
15. Tiri, B.; Sensi, E.; Marsiliani, V.; Cantarini, M.; Priante, G.; Vernelli, C.; Martella, L.A.; Costantini, M.; Mariottini, A Andreani, P.; et al. Antimicrobial Stewardship Program, COVID-19, and Infection Control: Spread of Carbapenem-Resista *Klebsiella Pneumoniae* Colonization in ICU COVID-19 Patients. What Did Not Work? *J. Clin. Med.* **2020**, *9*, 2744. [CrossRef]
16. Yang, X.; Liu, X.; Li, W.; Shi, L.; Zeng, Y.; Xia, H.; Huang, Q.; Li, J.; Li, X.; Hu, B.; et al. Epidemiological Characteristics ar Antimicrobial Resistance Changes of Carbapenem-Resistant *Klebsiella pneumoniae* and *Acinetobacter baumannii* under the COVID- Outbreak: An Interrupted Time Series Analysis in a Large Teaching Hospital. *Antibiotics* **2023**, *12*, 431. [CrossRef]
17. Pascale, R.; Bussini, L.; Gaibani, P.; Bovo, F.; Fornaro, G.; Lombardo, D.; Ambretti, S.; Pensalfine, G.; Appolloni, Bartoletti, M.; et al. Carbapenem-resistant bacteria in an intensive care unit during the coronavirus disease 2019 (COVID-1 pandemic: A multicenter before-and-after cross-sectional study. *Infect. Control Hosp. Epidemiol.* **2022**, *43*, 461–466. [CrossRef]

18. Barek, M.A.; Aziz, M.A.; Islam, M.S. Impact of age, sex, comorbidities and clinical symptoms on the severity of COVID-19 cases: A meta-analysis with 55 studies and 10,014 cases. *Heliyon* **2020**, *6*, e05684. [CrossRef]
19. Montrucchio, G.; Corcione, S.; Sales, G.; Curtoni, A.; De Rosa, F.; Brazzi, L. Carbapenem-resistant *Klebsiella pneumoniae* in ICU-admitted COVID-19 patients: Keep an eye on the ball. *J. Glob. Antimicrob. Resist.* **2020**, *23*, 398–400. [CrossRef]
20. Liu, P.; Li, X.; Luo, M.; Xu, X.; Su, K.; Chen, S.; Qing, Y.; Li, Y.; Qiu, J. Risk Factors for Carbapenem-Resistant *Klebsiella pneumoniae* Infection: A Meta-Analysis. *Microb. Drug Resist.* **2018**, *24*, 190–198. [CrossRef]
21. Mędrzycka-Dąbrowska, W.; Lange, S.; Zorena, K.; Dąbrowski, S.; Ozga, D.; Tomaszek, L. Carbapenem-Resistant *Klebsiella pneumoniae* Infections in ICU COVID-19 Patients-A Scoping Review. *J. Clin. Med.* **2021**, *10*, 2067. [CrossRef] [PubMed]
22. Pintado, V.; Ruiz-Garbajosa, P.; Escudero-Sanchez, R.; Gioia, F.; Herrera, S.; Vizcarra, P.; Fortún, J.; Cobo, J.; Martín-Dávila, P.; Morosini, M.I.; et al. Carbapenemase-producing *Enterobacterales* infections in COVID-19 patients. *Infect. Dis.* **2022**, *54*, 36–45. [CrossRef] [PubMed]
23. Dumitru, I.M.; Dumitrascu, M.; Vlad, N.D.; Cernat, R.C.; Ilie-Serban, C.; Hangan, A.; Slujitoru, R.E.; Gherghina, A.; Mitroi-Maxim, C.; Curtali, L.; et al. Carbapenem-Resistant *Klebsiella pneumoniae* Associated with COVID-19. *Antibiotics* **2021**, *10*, 561. [CrossRef] [PubMed]
24. Gorrie, C.L.; Mirčeta, M.; Wick, R.R.; Edwards, D.J.; Thomson, N.R.; Strugnell, R.A.; Pratt, N.F.; Garlick, J.S.; Watson, K.M.; Pilcher, D.V.; et al. Gastrointestinal Carriage Is a Major Reservoir of *Klebsiella pneumoniae* Infection in Intensive Care Patients. *Clin. Infect. Dis.* **2017**, *65*, 208–215. [CrossRef]
25. Messika, J.; La Combe, B.; Ricard, J.D. Oropharyngeal colonization: Epidemiology, treatment and ventilator-associated pneumonia prevention. *Ann. Transl. Med.* **2018**, *6*, 426. [CrossRef]
26. Liao, W.; Liu, Y.; Zhang, W. Virulence evolution, molecular mechanisms of resistance and prevalence of ST11 carbapenem-resistant *Klebsiella pneumoniae* in China: A review over the last 10 years. *J. Glob. Antimicrob. Resist.* **2020**, *23*, 174–180. [CrossRef]
27. Wang, J.; Feng, Y.; Zong, Z. Worldwide transmission of ST11-KL64 carbapenem-resistant *Klebsiella pneumoniae*: An analysis of publicly available genomes. *mSphere* **2023**, e0017323. [CrossRef]
28. Xie, M.; Yang, X.; Xu, Q.; Ye, L.; Chen, K.; Zheng, Z.; Dong, N.; Sun, Q.; Shu, L.; Gu, D.; et al. Clinical evolution of ST11 carbapenem resistant and hypervirulent *Klebsiella pneumoniae*. *Commun. Biol.* **2021**, *4*, 650. [CrossRef]
29. Xu, M.; Qian, C.; Jia, H.; Feng, L.; Shi, S.; Zhang, Y.; Wang, L.; Cao, J.; Zhou, T.; Zhou, C. Emergence of Ceftazidime-Avibactam Resistance and Decreased Virulence in Carbapenem-Resistant ST11 *Klebsiella pneumoniae* During Antibiotics Treatment. *Infect. Drug Resist.* **2022**, *15*, 6881–6891. [CrossRef]
30. Heiden, S.E.; Hübner, N.-O.; Bohnert, J.A.; Heidecke, C.-D.; Kramer, A.; Balau, V.; Gierer, W.; Schaefer, S.; Eckmanns, T.; Gatermann, S.; et al. A *Klebsiella pneumoniae* ST307 outbreak clone from Germany demonstrates features of extensive drug resistance, hypermucoviscosity, and enhanced iron acquisition. *Genome Med.* **2020**, *12*, 113. [CrossRef]
31. Fritzenwanker, M.; Imirzalioglu, C.; Herold, S.; Wagenlehner, F.M.; Zimmer, K.P.; Chakraborty, T. Treatment Options for Carbapenem-Resistant Gram-Negative Infections. *Dtsch. Arztebl. Int.* **2018**, *115*, 345–352. [CrossRef]
32. Zhu, J.; Chen, Y.; Yang, X. Antibiotic Resistance, Molecular Characteristics and Risk Factors of Carbapenem-Resistant *Klebsiella pneumoniae* in Clinical Isolates. *Infect. Drug Resist.* **2022**, *15*, 6671–6680. [CrossRef]
33. Rodríguez-Baño, J.; Gutiérrez-Gutiérrez, B.; Machuca, I.; Pascual, A. Treatment of Infections Caused by Extended-Spectrum-Beta-Lactamase-, AmpC-, and Carbapenemase-Producing *Enterobacteriaceae*. *Clin. Microbiol. Rev.* **2018**, *31*, e00079-17. [CrossRef]
34. Karampatakis, T.; Tsergouli, K.; Behzadi, P. Carbapenem-Resistant *Klebsiella pneumoniae*: Virulence Factors, Molecular Epidemiology and Latest Updates in Treatment Options. *Antibiotics* **2023**, *12*, 234. [CrossRef] [PubMed]
35. El-Sayed Ahmed, M.A.E.G.; Zhong, L.-L.; Shen, C.; Yang, Y.; Doi, Y.; Tian, G.-B. Colistin and its role in the Era of antibiotic resistance: An extended review (2000–2019). *Emerg. Microbes Infect.* **2020**, *9*, 868–885. [CrossRef] [PubMed]
36. Torres, D.A.; Seth-Smith, H.M.B.; Joosse, N.; Lang, C.; Dubuis, O.; Nüesch-Inderbinen, M.; Hinic, V.; Egli, A. Colistin resistance in Gram-negative bacteria analysed by five phenotypic assays and inference of the underlying genomic mechanisms. *BMC Microbiol.* **2021**, *21*, 321. [CrossRef] [PubMed]
37. Yu, L.; Zhang, J.; Fu, Y.; Zhao, Y.; Wang, Y.; Zhao, J.; Guo, Y.; Li, C.; Zhang, X. Synergetic Effects of Combined Treatment of Colistin With Meropenem or Amikacin on Carbapenem-Resistant *Klebsiella pneumoniae* in vitro. *Front. Cell. Infect. Microbiol.* **2019**, *9*, 422. [CrossRef] [PubMed]
38. Paul, M.; Daikos, G.L.; Durante-Mangoni, E.; Yahav, D.; Carmeli, Y.; Benattar, Y.D.; Skiada, A.; Andini, R.; Eliakim-Raz, N.; Nutman, A.; et al. Colistin alone versus colistin plus meropenem for treatment of severe infections caused by carbapenem-resistant Gram-negative bacteria: An open-label, randomised controlled trial. *Lancet Infect. Dis.* **2018**, *18*, 391–400. [CrossRef] [PubMed]
39. Geng, T.T.; Xu, X.; Huang, M. High-dose tigecycline for the treatment of nosocomial carbapenem-resistant *Klebsiella pneumoniae* bloodstream infections: A retrospective cohort study. *Medicine* **2018**, *97*, e9961. [CrossRef] [PubMed]
40. Jin, X.; Chen, Q.; Shen, F.; Jiang, Y.; Wu, X.; Hua, X.; Fu, Y.; Yu, Y. Resistance evolution of hypervirulent carbapenem-resistant *Klebsiella pneumoniae* ST11 during treatment with tigecycline and polymyxin. *Emerg. Microbes Infect.* **2021**, *10*, 1129–1136. [CrossRef]
41. Bulik, C.C.; Nicolau, D.P. Double-carbapenem therapy for carbapenemase-producing *Klebsiella pneumoniae*. *Antimicrob. Agents Chemother.* **2011**, *55*, 3002–3004. [CrossRef]

42. Tan, X.; Kim, H.S.; Baugh, K.; Huang, Y.; Kadiyala, N.; Wences, M.; Singh, N.; Wenzler, E.; Bulman, Z.P. Therapeutic Options fo Metallo-beta-Lactamase-Producing Enterobacterales. *Infect. Drug Resist.* **2021**, *14*, 125–142. [CrossRef]
43. Marshall, S.; Hujer, A.M.; Rojas, L.J.; Papp-Wallace, K.M.; Humphries, R.M.; Spellberg, B.; Hujer, K.M.; Marshall, E.K.; Rudin, S.D Perez, F.; et al. Can Ceftazidime-Avibactam and Aztreonam Overcome beta-Lactam Resistance Conferred by Metallo-beta Lactamases in *Enterobacteriaceae*? *Antimicrob. Agents Chemother.* **2017**, *61*, e02243-16. [CrossRef]
44. Mauri, C.; Maraolo, A.E.; Di Bella, S.; Luzzaro, F.; Principe, L. The Revival of Aztreonam in Combination with Avibactam agains Metallo-beta-Lactamase-Producing Gram-Negatives: A Systematic Review of In Vitro Studies and Clinical Cases. *Antibiotics* **2021** *10*, 1012. [CrossRef] [PubMed]
45. Spaziante, M.; Oliva, A.; Ceccarelli, G.; Venditti, M. What are the treatment options for resistant *Klebsiella pneumoniae* carbapene mase (KPC)-producing bacteria? *Expert Opin. Pharmacother.* **2020**, *21*, 1781–1787. [CrossRef]
46. Jean, S.S.; Harnod, D.; Hsueh, P.R. Global Threat of Carbapenem-Resistant Gram-Negative Bacteria. *Front. Cell. Infect. Microbio* **2022**, *12*, 823684. [CrossRef] [PubMed]
47. Wu, N.; Chen, L.K.; Zhu, T. Phage therapy for secondary bacterial infections with COVID-19. *Curr. Opin. Virol.* **2022**, *52*, 9–14 [CrossRef] [PubMed]
48. Zurabov, F.M.; Chernevskaya, E.A.; Beloborodova, N.V.; Zurabov, A.Y.; Petrova, M.V.; Yadgarov, M.Y.; Popova, V.M Fatuev, O.E.; Zakharchenko, V.E.; Gurkova, M.M.; et al. Bacteriophage Cocktails in the Post-COVID Rehabilitation. *Viruses* **2022** *14*, 2614. [CrossRef]
49. FDA Allows Phage Therapy for COVID19 Patients. Available online: https://www.bacteriophage.news/fda-phage-therapy-for covid19-patients/ (accessed on 27 June 2023).
50. Loc-Carrillo, C.; Abedon, S.T. Pros and cons of phage therapy. *Bacteriophage* **2011**, *1*, 111–114. [CrossRef]
51. Gattringer, R.; Nikš, M.; Ostertág, R.; Schwarz, K.; Medvedovic, H.; Graninger, W.; Georgopoulos, A. Evaluation of MIDITECH automated colorimetric MIC reading for antimicrobial susceptibility testing. *J. Antimicrob. Chemother.* **2002**, *49*, 651–659. [CrossRef]
52. EUCAST Breakpoint Tables for Interpretation oh MICs and Zone Diameters. Version 13.0.2023. Available online: https://www.eucast. org/fileadmin/src/media/PDFs/EUCAST_files/Breakpoint_tables/v_13.0_Breakpoint_Tables.pdf (accessed on 27 June 2023).
53. EUCAST Guidelines for Detection of Resistance Mechanisms and Specific Resistances of Clinical and/or Epidemiological Impor tance, Version 2.01. 2017. Available online: https://www.eucast.org/fileadmin/src/media/PDFs/EUCAST_files/Resistance_ mechanisms/EUCAST_detection_of_resistance_mechanisms_170711.pdf (accessed on 27 June 2023).
54. Prjibelski, A.; Antipov, D.; Meleshko, D.; Lapidus, A.; Korobeynikov, A. Using SPAdes De Novo Assembler. *Curr. Protoc. Bioinform.* **2020**, *70*, e102. [CrossRef] [PubMed]

antibiotics

MDPI

rticle

Iigh Emergence of Multidrug-Resistant Sequence Type .31 Subclade C2 among Extended-Spectrum β-Lactamase ESBL)-Producing *Escherichia coli* Isolated from the Jniversity Hospital Bratislava, Slovakia

ín Koreň [1], Michal Andrezál [2], Elham Ozaee [2], Hana Drahovská [2], Martin Wawruch [3], Adriána Liptáková [1,*] and ibor Maliar [4]

[1] Institute of Microbiology, Faculty of Medicine, Comenius University, University Hospital Bratislava, 81108 Bratislava, Slovakia; jan.koren@fmed.uniba.sk

[2] Department of Molecular Biology, Faculty of Natural Sciences, Comenius University, 84215 Bratislava, Slovakia; michal.andrezal@uniba.sk (M.A.); elham.ozaee@uniba.sk (E.O.); hana.drahovska@uniba.sk (H.D.)

[3] Institute of Pharmacology and Clinical Pharmacology, Faculty of Medicine, Comenius University, 81108 Bratislava, Slovakia; martin.wawruch@fmed.uniba.sk

[4] Department of Chemistry, Faculty of Natural Sciences, University of Ss. Cyril and Methodius in Trnava, 91701 Trnava, Slovakia; tibor.maliar@ucm.sk

* Correspondence: adriana.liptakova@fmed.uniba.sk; Tel.: +421-0908-962-530

Abstract: The expansion of sequence type 131 (ST131) extended-spectrum β-lactamase (ESBL)-producing *Escherichia coli* (*E. coli*) represents major worldwide challenges. *E. coli* strains originating from healthcare facilities (labeled No. 1 and No. 2) of the University Hospital Bratislava (UHB) were analyzed for ST131 emergence, including its (sub)lineages and clonal relatedness. Antimicrobial resistance was determined in most strains. Of a total of 354 *E. coli* strains, 263 (74.3%) belonged to ST131; of these, 177 (67.3%) were from No. 1. Generally, among 260 ST131 *E. coli*, clades A/B were confirmed in 20 (7.7%), while clade C was noted in 240 (92.3%) strains; within them, subclades were detected as follows: C0 (17; 7.1%), C1 (3; 1.2%), and C2 (220; 91.7%). Among fifteen randomly selected *E. coli* strains that were investigated for ST and clonal relatedness, seven STs were identified: eight (53.3%) ST131, two (13.3%) ST73, and one each (6.7%) of ST10, ST12, ST14, ST1193, and ST1196. From No. 1, two ST131 in the first internal clinic and one ST131 from No. 2 in the aftercare department were highly clonally related, suggesting possible epidemiological association. Antimicrobial resistance was as follows: ciprofloxacin 93.8%, ceftazidime 78.4%, meropenem 0%, fosfomycin 2.9% and nitrofurantoin 1.4%. Prevention of ESBL-producing *E. coli* dissemination, especially for ST131 clade C2, is inevitably necessary for reducing drug resistance and decreasing healthcare-associated infections.

Keywords: antimicrobial resistance; clade and subclade; ESBL-producing *Escherichia coli*; healthcare facilities; ST131

1. Introduction

The global emergence of multidrug resistance (MDR) to extended-spectrum β-lactamase (ESBL)-producing strains represents an ongoing rising threat of healthcare-associated infections contributing to the antimicrobial resistance crisis. The potential driver of this might be ineffective early empirical therapy resulting in the overuse of reserve drugs required for difficult-to-treat pathogens [1,2]. In addition, the elevation in MDR strains is correspondingly related to the overall increase in the administration of antimicrobial agents, which affects the composition of the microbiota in humans [3]. *Escherichia coli* (*E. coli*) strains have been recognized as some of the most prevalent ESBL-producing Gram-negative pathogens [4–6] from the *Enterobacterales* order. The prevalence of ESBL-producing

E. coli strains has increased in clinical samples to more than 10% in 22 European countries in previous years. Increased colonization in the gastrointestinal tract may have contributed to this rise even in countries that until now had a low proportion of ESBL strains, such as Sweden and the Netherlands [7]. Recent molecular epidemiologic studies have confirmed the increasing tendency for both hospital-acquired and community-acquired extraintestinal infections, especially urinary tract and bloodstream infections caused by ESBL-producing *E. coli* with sequence type (ST) 131, which was more frequent in human samples than other STs [6,8]. In particular, urinary tract infections caused by ST131 were more often persistent, had an unfavorable treatment outcome, and were related to urinary catheterization, diabetes mellitus, immunodeficiency, and higher age [1]. On the other hand, *E. coli* strains were the leading cause of bloodstream infections, and ESBL production was recorded up to 17%, while ST131 was reported to be one of the most frequent clones implicated in colonization, hospital transmission, morbidity, and mortality [9]. Accordingly, we can emphasize that these mostly extraintestinal pathogenic *E. coli* (ExPEC, from extraintestinal sites) strains are concerning for public health due to being important etiological agents, and they are most often associated with the pandemic clonal group denoted as ST131 that has resistance to fluoroquinolones and third-generation cephalosporins [10,11]. Nowadays, ST131 is associated with many MDR infections in developed and developing countries [1]. The occurrence of ESBL-producing *E. coli* strains was recorded in one-third (34.2%) of clinical samples in recent years in Slovakia [12], but research regarding the emergence of high-risk clones such as, e.g., ST131 or others, has not yet been published from our country, as they were confirmed in the former common state of the Czech Republic [13].

Expansion of the ST131 clone is commonly related to plasmid-disseminated $bla_{\text{CTX-M}}$ genes, including genetic determinants of fluoroquinolone resistance. The H30 subclone of ST131 is defined according to its allele 30 of *fimH* (fimbrial adhesin-encoding gene type 1), and this genetic form has become the most prevalent [14–16]. Other *fimH* allele variants of the ST131 clone correspond to distinct major lineages or clades: A/H41, B/H22, and C/H30. The main generator of antimicrobial resistance is *E. coli* from the predominant clade C, mostly comprising subclades C1 and C2 (formerly H30R and H30Rx) [10,17,18]. The latter, subclade C2, is mainly associated with $bla_{\text{CTX-M-15}}$ and sublineage C1, which can be related to $bla_{\text{CTX-M-27}}$ (cluster C1-M27) or not (cluster non-C1-M27). The variant C0 is regarded as the progenitor of subclades C1 and C2 [10,19].

The objective of this study was to investigate the prevalence of ST131 among ESBL-producing *Escherichia coli* strains in two healthcare facilities within the University Hospital Bratislava (UHB). Further, the representation of ST131 lineages or clades A, B, or C was determined, and within the last (clade C), subclades C0, C1, and C2 were detected using the multiplex PCR method. Selected ESBL- and/or AmpC β-lactamase-producing *E. coli* strains, including nonproducers from both healthcare facilities, were also explored for genetic relatedness by the next-generation sequencing method. In addition, susceptibility to antimicrobial agents was ascertained in the examined strains. The resistance profile was evaluated and used to statistically compare the resistance rate associated with ST131 and non-ST131 in mostly ESBL-positive *E. coli* strains. Finally, binary logistic regression was used to examine patient- and *E. coli*-related characteristics associated with increased antimicrobial resistance.

2. Results

2.1. Patients' Characteristics and Isolation of ESBL-Producing E. coli Strains

In total, 354 ESBL- and/or AmpC β-lactamase-producing *E. coli* strains were isolated from patients residing in two healthcare facilities within the UHB. The studied patient cohort comprised 235 (66.4%) females, and their mean age was 80.4 ± 11.9 (range from 9 to 98) years. According to the Mann–Whitney U test, the difference when comparing both genders, including age, was statistically significant ($p < 0.001$). Overall, out of all patients, an age of 65 or over was predominant in 310 (87.6%) individuals, as shown in Table 1. Healthcare facility No. 1 had a larger number of ESBL-producing *E. coli* strains, with 243 (68.6%) originating here; most of them, that is, 190 (53.7%), were from the first internal clinic. In healthcare setting No. 2, 111 (31.4%) examined strains were isolated; a majority of these, 76 (21.5%), were mainly from the geriatric clinic. Regarding the type of specimen, the urine sample represented the largest number at 227 (64.1%). Specimens collected from the skin and soft tissue were less frequent, with 50 (14.1%) being found, and they were associated with a wound, ulcus cruris, abscess, fistula, skin lesion or defect, or pattern (lungs and spleen swab) obtained via the postmortem of dead patients from the Institute of Pathological Anatomy. There were 29 (8.2%) hemocultures obtained. From the genital tract, two (0.6) samples, one a vaginal swab and one ejaculate, were collected. In the cases of catheter-related infections, the four (1.1%) tips of the catheters mainly originated from the surgical site of infection. Six (1.7%) sputum samples were collected. Finally, 22 (6.2%) throat or nasal swabs and 14 (4%) stools or rectal swabs were sampled. According to defined criteria conditions, including a significant microbiological finding, our set of evaluated clinical strains was associated with 287 (81.1%) cases of infection and 67 (18.9%) cases of colonization.

Table 1. Patients' characteristics including the occurrence of ST131 and non-ST131 *Escherichia coli* extended-spectrum β-lactamase-producing strains in patients and healthcare settings.

		Total of (354)		ST131 (263)		Non-ST131 (91)	
Gender and Age		*n*	%	*n*	%	*n*	%
Male Mean years (range) 75.0 ± 14.4 (25–96) Median years 79		119	33.6	86	32.7	33	36.3
Female Mean years (range) **80.4** ± 11.9 (9–98) * Median years 83		235	66.4	177	67.3	58	63.7
≥65 years		310	87.6	236	**89.7** **	74	81.3
<65 years		44	12.4	27	10.3	17	18.7
Setting							
No. 1	First internal clinic	190	53.7	142	54.0	48	52.7
	Surgical clinic	20	5.6	13	4.9	7	7.7
	Neurological clinic	7	2.0	6	2.3	1	1.1
	Psychiatric clinic	5	1.4	5	1.9	0	0
	Dermatovenerological clinic	8	2.3	6	2.3	2	2.2
	Urological outpatient clinic	10	2.8	5	1.9	5	5.5
	Institute of Pathological Anatomy	3	0.8	0	0	3	3.3
No. 2	Geriatric clinic	76	21.5	58	22.1	18	19.8
	Long-term care department	18	5.1	14	5.3	4	4.4
	Aftercare department	12	3.4	11	4.2	1	1.1
	Other outpatient offices	5	1.4	3	1.1	2	2.2

n—number of patients, cases, or specimens; %—percentage; **No. 1** and **No. 2**—designations of selected healthcare facilities that belong to the University hospital Bratislava (UHB); *—$p < 0.001$; and **—$p = 0.036$.

2.2. Occurrence of ST131 and non-ST131 ESBL-Producing E. coli Strains

A total of 354 multidrug-resistant and mostly ESBL-producing *E. coli* strains were investigated for the presence of ST131. Suspected pathogens were subjected to ST131 clade multiplex PCR with *fimH* detection [10]. Overall, 263 (74.3%) ST131 and 91 (25.7%) non-ST131 mostly ESBL-producing *E. coli* strains were obtained. ST representation in males was as follows: 86 (32.7%) ST131 strains and 33 (36.3%) non-ST131 strains. Conversely, females had a higher number of ST131 strains at 177 (67.3%) than non-ST131 strains at 58 (63.7), without statistical significance, as shown in Table 1. In addition, an increased prevalence of 236 (89.7%) ST131 strains in comparison with 74 (81.3%) non-ST131 strains was confirmed in patients aged 65 years and older with statistical significance ($p = 0.036$). Considering healthcare facility No. 1, the proportion of STs was as follows: 177 (67.3%) ST131 strains and 66 (72.5%) non-ST131 strains, but within some clinics, e.g., the first internal clinic and neurological clinic, a slightly higher rate of ST131 versus (vs.) non-ST131 was recorded, as shown in Table 1. On the other hand, in healthcare facility No. 2, a higher number of ST131 strains compared to non-ST131 were detected, being 86 (32.7%) vs. 25 (27.5%), respectively, and also within individual workplaces, such as the geriatric clinic, long-term care department, and aftercare department, a raised rate of ST131 compared to non-ST131 was found, as presented in Table 1. However, it was not statistically significant. Regarding the specimens, ST131 was more common than non-ST131, e.g., in urine samples (172; 65.4% vs. 55; 60.4%), blood cultures (25; 9.5% vs. 4; 4.4%), and throat or nasal swabs (17; 6.5% vs. 5; 5.5%).

2.3. *Distribution of Clades and Subclades of ST131 ESBL-Producing E. coli Strains*

Out of 263 ST131 ESBL-positive *E. coli* strains, 260 were used to determine lineages or clades A and B (A/B) and C, including subclades (C0, C1, and C2). Three ST131 strains remained undetermined. Clades A/B were recorded 20 times (7.7%), while clade C was determined for 240 (92.3%) ST131 *E. coli* strains. Within clade C, 17 (7.1%) cases of subclade C0 were identified, and there were 3 (1.2%) strains of C1 and 220 (91.7%) pathogens belonging to subclade C2. Regarding both genders, clades A/B had the same distribution (10; 50%), but the following subclades had a higher representation in females: subclade C0 (13; 76.5%) and subclade C2 (152; 69.1%), while a higher rate of subclade C1 was noted in males (2; 66.7%), as indicated in Table 2. The prevalence of identified clades and subclades in patients aged 65 and over was higher—A/B (16; 80%), C0 (16; 94.1%), C1 (3; 100%), and C2 (200; 90.9%)—than in younger patients. From the viewpoint of healthcare facility No. 1, and mainly from the first internal clinic, there were 12 (60%) detected strains of clades A/B and the same number, 12 (70.6%), of subclade C0. However, most of the 116 (52.7%) ST131 ESBL-producing *E. coli* strains were from subclade C2 at the first internal clinic. In healthcare facility No. 2, particularly at the geriatric clinic, four (20%) cases of clades A/B, three (17.6%) cases of subclade C0, and two (66.7%) C1 occurrences were determined. However, the highest number of *E. coli* strains, 49 (22.3%), were identified to be subclade C2 within the same geriatric clinic; further data are presented in Table 2. Based on the urine samples, 15 (75%) ST131 ESBL-positive *E. coli* strains were determined as clades A/B, and 11 (64.7%) agents were identified as subclade C0; however, most cases, 142 (64.5%), were found to be subclade C2.

2.4. *Genetic Relatedness of ESBL-Producing Sequenced E. coli Strains*

Fifteen *E. coli* genomes were sequenced using next-generation sequencing (NGS) Illumina technology. High-quality contigs were obtained for all of the sequenced strains, with an average of 116 contigs and an average coverage of 77 times per sequenced genome. The length of genomes ranged from 3.7 Mbp to 5.3 Mbp.

The clonality of the strains was confirmed using MLST, which was used to determine the occurrence of seven sequence types (STs). Most samples (53.3%) belonged to ST131, followed by ST73 (13.3%). The remaining samples were assigned to ST10, ST12, ST14, ST1193, and ST1196 (each 6.7%).

The mutual relatedness of strains assigned to the same ST was later evaluated using core genome MLST (cgMLST). Strains belonging to ST131 showed high variability, with up to 659 different alleles between them (Figure 1). According to this analysis, the ST131 strains created seven unique profiles. Four strains (KMB-999, KMB-981, KMB-979, and KMB-984) differed only in 0–13 alleles and were clustered in one group. The rest of the ST131 strains were clonally unrelated and quite distinct, except for two ST131 strains (KMB-1000 and KMB-1002) from the urological outpatient clinic with high clonal relatedness. Seven remaining STs formed distant and unrelated forms, including the two ST73 strains, which mutually differed in 227 alleles (Figure 1).

Table 2. Distribution of clades and subclades of ST131 *Escherichia coli* extended-spectrum β-lactamase-producing strains relating to patients and healthcare settings.

		Total of (260)		A/B (20)		C0 (17)		C1 (3)		C2 (220)	
Gender		*n*	%	*n*	%	*n*	%	*n*	%	*n*	%
Male		84	32.3	10	50	4	23.5	2	66.7	68	30.9
Female		176	67.7	10	50	13	76.5	1	33.3	152	69.1
Age											
≥65 years		235	90.4	16	80	16	94.1	3	100	200	90.9
<65 years		25	9.6	4	20	1	5.9			20	9.1
Setting											
No. 1	First internal clinic	140	53.8	12	60	12	70.6			116	52.7
	Surgical clinic	13	5	1	5					12	5.4
	Neurological clinic	6	2.3							6	2.7
	Psychiatric clinic	5	1.9							5	2.3
	Dermatovenerological clinic	6	2.3							6	2.7
	Urological outpatient clinic	5	1.9	2	10					3	1.4
	Institute of Pathological Anatomy										
No. 2	Geriatric clinic	58	22.3	4	20	3	17.6	2	66.7	49	22.3
	Long-term care department	14	5.4			2	11.8			12	5.5
	Aftercare department	10	3.8	1	5			1	33.3	8	3.6
	Other outpatient offices	3	1.2							3	1.4

A/B—clades A and B; **C0**, **C1**, and **C2**— subclades; *n*—number of patients, cases, or specimens; %—percentage; and **No. 1** and **No. 2**—designations of selected healthcare facilities that belong to the University Hospital Bratislava (UHB).

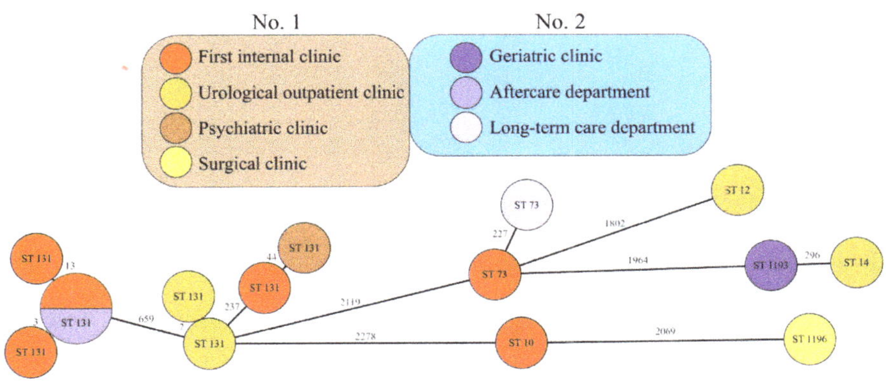

Figure 1. GrapeTree of 15 assayed *E. coli* strains based on cgMLST. Circle sizes are consistent with the count of strains with the same sequence types (STs) noted inside the circle. Numbers on lines between circles indicate the values of distinct alleles between groups. The colors represent the origin of the strains—clinics or departments, as well as healthcare facilities designated by healthcare setting number (**No. 1** or **No. 2**)—including clonal relatedness.

2.5. Occurrence of STs in Healthcare Facilities

Fifteen randomly selected *E. coli* strains originating from two healthcare facilities following sequencing were assigned to STs. A total of 12 strains belonged to healthcare facility No. 1, of which the most ST131 strains (4) were from the first internal clinic, including 1 ST10 and 1 ST73 strain. From healthcare facility No. 2, three *E. coli* strains belonged to different STs (ST73, ST131, and ST1193). They originated from various hospital departments, including a clinic, as illustrated in Figure 1.

2.6. Antibiotic Resistance of MDR and ESBL-Producing E. coli Strains

Out of a total set of 354 *E. coli* strains, the presence of $bla_{CTX\text{-}M}$ genes was investigated; 350 (98.9%) were found to be positive for the production of CTX-M extended-spectrum β-lactamase, without specifying the variant in more detail. Using the sequencing method, the production of AmpC β-lactamase was encountered in 11 (3.1%) *E. coli* strains and 7 (2%) possible causative agents were found to concurrently produce ESBL, including AmpC β-lactamase. Four (1.1%) $bla_{CTX\text{-}M}$-negative strains only carried out AmpC production. One *E. coli* strain produced KPC (*Klebsiella pneumoniae* carbapenemase) from serine group A. The bla_{KPC} gene encoding this carbapenemase was confirmed using the PCR method at the National Reference Centre (NRC) for monitoring antimicrobial resistance in the Slovak Republic.

The susceptibility of MDR *E. coli* with ESBL production was evaluated, and the resistance rate to antimicrobial agents was compared between ST131 and non-ST131 *E. coli* strains. Resistance to ST131 vs. non-ST131 was determined to be statistically significant as follows: 79.1% vs. 67.5% to ampicillin/sulbactam (p = 0.037), 20.3% vs. 14.5% to piperacillin/tazobactam (p = 0.002), 99.2% vs. 96.1% to cefotaxime (p = 0.043), and 81.6% vs. 68.4% to ceftazidime (p = 0.042). Overall, resistance to ceftazidime/avibactam and meropenem was not confirmed, while resistance to ertapenem was recorded at 0.5% in ST131 vs. 3% in non-ST131 *E. coli* strains, and further resistance to gentamicin was found at 22.3% vs. 31.2%, respectively; however, these values were not statistically significant. Resistance to amikacin was detected in ST131 at 4.1% vs. non-ST131 *E. coli* strains at 5.3% (p = 0.038), and ciprofloxacin resistance was recorded at 96.7% vs. 84.4% (p < 0.001), respectively, with statistical significance; these data are listed in Table 3, including data on resistance to other antimicrobial agents.

Table 3. Antimicrobial resistance of total, ST131, and non-ST131 *Escherichia coli* ESBL-producing strains.

Antibiotics	Total			ST131 Strains			Non-ST131 Strains			p *
	n	R	(%)	*n*	R	(%)	*n*	R	(%)	
Ampicillin/sulbactam	321	245	76.3	244	193	**79.1**	77	52	67.5	0.037
Piperacillin/tazobactam	317	60	18.9	241	49	**20.3**	76	11	14.5	0.002
Cefuroxime	320	317	99.1	243	242	99.6	77	75	97.4	
Cefotaxime	319	314	98.4	243	241	**99.2**	76	73	96.1	0.043
Ceftazidime	320	251	78.4	244	199	**81.6**	76	52	68.4	0.042
Ceftazidime/avibactam	347	0	0	262	0	0	85	0	0	
Cefoperazone/sulbactam	280	18	6.4	214	15	7	66	3	4.5	
Cefepime	281	234	83.3	214	182	85	67	52	77.6	
Ertapenem	280	3	1.1	214	1	0.5	66	2	3	
Meropenem	319	0	0	243	0	0	76	0	0	
Gentamicin	319	78	24.5	242	54	22.3	77	24	31.2	
Tobramycin	280	154	55	214	121	56.5	66	33	50	
Amikacin	320	14	4.4	244	10	4.1	76	4	**5.3**	0.038
Tetracycline	299	190	63.5	226	145	64.2	73	45	61.6	
Tigecycline	280	3	1.1	214	3	1.4	66	0	0	
Ciprofloxacin	321	301	93.8	244	236	**96.7**	77	65	84.4	<0.001
Colistin	276	1	0.4	211	1	0.5	65	0	0	
Trimethoprim/ sulfamethoxazole	319	241	75.5	242	186	76.9	77	55	71.4	
Nitrofurantoin	72	1	1.4	51	1	2	21	0	0	
Fosfomycin	69	2	2.9	49	1	2	20	1	5	

n—number of strains; **R**—*n* of resistant strains; %—percentage; and *—p < 0.05 values were statistically significant in comparisons between ST131 and non-ST131 resistance rate (%), which is indicated by boldface text.

In addition, the ability to predict patients' characteristics was evaluated, and *E. coli* clades and subclades were analyzed based on the emergence of antimicrobial resistance. It was found that the male gender was associated with gentamicin resistance (p = 0.049), with an odds ratio (OR) of 2.029 (95% CI [confidence interval], 1.003–4.104) with statistical significance. Similarly, statistically significant relations were represented by a specimen collected from the upper respiratory tract ($p < 0.001$; OR 0.032; 95% CI, 0.06–0.159); collection from the medical setting of the urological outpatient clinic ($p = 0.048$; OR 0.09; 95% CI, 0.008–0.977); and the presence of ST131 subclade C2 associated with resistance to ciprofloxacin ($p = 0.038$; OR 6.238; 95% CI, 1.106–35.176). Resistance to ceftazidime was also predicted by a specimen collected from the upper respiratory tract ($p = 0.001$; OR 0.102; 95% CI, 0.026–0.394); by the type of specimen being sputum ($p = 0.012$; OR 0.047; 95% CI, 0.004–0.515); and also by ST131 subclade C2 ($p = 0.001$; OR 8.41; 95% CI, 2.534–27.918) with statistical significance.

3. Discussion

Carbapenemase-producing *E. coli* ST131 strains have been spreading and causing diseases in recent years [20], and the treatment options are limited for these extensively drug-resistant (XDR, i.e., susceptible to one or two antimicrobial classes) pathogens. On the other hand, the dissemination of a pandemic high-risk clone designated as multidrug-resistant *E. coli* ST131 is worrying, as it could lead to an emergency in the public healthcare system due to the production of ESBL with resistance to third-generation cephalosporins and other classes of antibiotics [1,10]. The drugs of choice for these ESBL-producing *E. coli* pathogens causing severe and life-threatening infections are carbapenems, but the increasing administration of these drugs, which are considered last-resort antimicrobial agents, can result in the emergence of carbapenem resistance [4].

In studies conducted in Denmark [21] and Sweden [18], in which ESBL-producing *E. coli* was isolated from various biological samples or urine, respectively, the findings were comparable to our research results. Even though the proportion of females in these surveys was almost the same (65—66%) as in our records, the median age was 14 years lower (69 years), and the proportion of ST131 strains (38%) was lower [21] compared to our data (Table 1). Lindblom et al. (2022) [18] reported a much lower median age of 65 for inpatient females in a previous examination compared to that in our survey (Table 1), in which the overall proportion of patients in the age category of 65 years and older was seven times higher than the number of younger individuals. Our results confirm the occurrence of ST131 in older patients, while the prevalence of this clone may differ according to regions and countries.

In recent decades, many studies have emphasized the importance of the worldwide ST131 clone responsible for the spreading of antimicrobial-resistant infections with ESBL (mainly CTX-M β-lactamase) and coresistant profiles becoming dominant, including with regard to its acquisition in hospital [9] and community [22,23] settings. Subsequently, in the Lindblom et al. (2022) study, there was also a 2-fold higher ratio of patients from the community compared to inpatients [18]. In our group, there were a small number of patients from the urology outpatient clinic and other outpatient clinics (healthcare facilities No. 1 and No. 2, respectively). Regarding community carriage, which may contribute to infections, van den Bunt et al.'s (2020) study investigated the occurrence of ST131 in the gastrointestinal tract. They found that the most common risk factors were travelling related to prolonged carriage and antimicrobial administration [24]. In comparison, colonization was not detected in our outpatients.

A survey from the United States underlines that the ST131-H30 subclone (or clade C) is more common in hospitalized patients, in more at-risk elderly male patients in long-term care facilities (LTCF) [1]. On the contrary, compared to our results, ST131 clade C (together, C0, C1, and C2) was more represented in the female gender. In association with healthcare facility No. 2, it was more prevalent in the geriatric clinic than in the long-term care department (Table 2). In addition, in a study by Martischang et al. (2021), ST131 H30 or clade C was confirmed in a lower (68.6%) number of samples that originated from LTCF patients than in our survey. The reason for the lower occurrence of clade C could be that the prevalence of this subclone has had a downward tendency since 2015 [17]. Moreover, Flament-Simon et al. (2020) demonstrated the following data of ST131 clades and subclades of *E. coli* with ESBL production: clade A, 5.1%; subclade C1, 27.9%, and subclade C2, 65.8% [25]. According to our outcomes, clades A/B were detected with similarly low incidence, and the C1 subclade was detected in very few values. At the same time, subclade C2 showed in our survey an 18.8% higher prevalence (Table 2) than was found in the above-mentioned joint research from Spain and France. A variety of STs have been reported from hospital environments in this region [25], including ST10 and the newly emerging global ST1193, which were also confirmed in the first internal clinic (No. 1) and the geriatric clinic (No. 2), respectively, but so far without epidemiological, i.e., clonal, relatedness (Figure 1). Recently, this rapid and ongoing expansion of ST1193 was described in U.S. cities related to young patients less than 40 years old [26].

Two ST131 *E. coli* strains isolated from the first internal clinic and from the urological outpatient clinic (No. 1), together with one isolate from the aftercare department (No. 2), were highly clonally related, with only zero to three different alleles between each them (Figure 1); these findings suggest a possible epidemiological transmission within the healthcare facilities of No. 1, and between No. 1 and No. 2. The other randomly selected STs occurred sporadically, and they showed a marked difference both in STs and in allele diversity and did not appear to be epidemiologically significant or genetically related. From the largest surveillance research on multidrug-resistant pathogens, including *E. coli*, Roberts et al. (2022) published data on epidemiologic confirmations that support clustering, which is defined as the overlap of patients in the intensive care unit (ICU) with another in the same cluster. The number of confirmed overlaps for *E. coli* strains (11; 50%) was lower than other nosocomial strains. However, this high value represents half of the assessed *E. coli* strains [27]. From the given data, it can be pointed out that MDR *E. coli* strains are of great importance in the dissemination between patients. In small numbers, this was ascertained by our investigation.

In an observational cohort survey from Vietnam, the genomic characteristics of MDR *E. coli* were evaluated [27]. The resistance genes *bla*$_{\text{CTX-M}}$ (85%) were found to have the highest prevalence, while *bla*$_{\text{KPC}}$ (13%) and *bla*$_{\text{NDM}}$ (24%) occurred in lower but important numbers. Compared to our data, the same number of ST131 and ST1193 strains were also confirmed in our study, but *bla*$_{\text{CTX-M}}$ genes were detected in the whole set (except for four strains) with a much higher prevalence. A more precise detection of *bla*$_{\text{CTX-M}}$ genes was not performed. Based on subclades C1 and C2 (Table 2), the possible presence of *bla*$_{\text{CTX-M-27}}$ or not, and *bla*$_{\text{CTX-M-15}}$ [10,19,25], respectively, can be assumed. On the other hand, carbapenemase production was confirmed in only one strain by evidence of *bla*$_{\text{KPC}}$ genes, which is a remarkable difference compared to the study by Roberts et al. (2022), in which almost a quarter of *E. coli* were confirmed to produce metallo-carbapenemase NDM (New Delhi metallo-β-lactamase) [27]. In line with our knowledge from this investigation, it could be highlighted that the emergence of carbapenemase-producing *E. coli* at our monitored medical facilities so far has been sporadic, but not insignificant. Correspondingly, it is necessary to consider warnings about the possible rise in carbapenem resistance even in *E. coli* strains [20]. In addition, ST361 *E. coli* that produces two carbapenemases encoded *bla*$_{\text{KPC-3}}$ and *bla*$_{\text{NDM-5}}$ was detected in a patient hospitalized in Switzerland, and this strain was resistant even to the most recent generations of antimicrobial agents (cefiderocol and aztreonam/avibactam). At the same time, it was sensitive to older agents (tigecycline, fosfomycin, and colistin) [28]. The situation is similar with colistin resistance. In our outcomes, only one ST131 *E. coli* strain was resistant to this last-line resort antibiotic (Table 3). In the findings of Roberts et al. (2022), resistance to colistin (4%) mediated by *mcr* genes (mediated by the colistin resistance gene) is reported with a slightly higher incidence [27]. It is crucial to be aware that although colistin-resistant strains are rare in our healthcare-associated facilities, if such a threat emerges worldwide, it may unexpectedly extend to our medical settings.

MacFadden et al. (2019) published results related to *E. coli* bloodstream infections, and they reported that inappropriate empiric antimicrobial therapy increases mortality [29]. In our data, the occurrence of *E. coli* blood infections with ESBL production was noted, of which, out of 29 patients, 25 cases were caused by ST131, and 21 of these cases were caused by subclade C2, which is mostly *bla*$_{CTX-M-15}$ positive. Concerning our hemoculture outcomes, empiric treatment for susceptible *E. coli* may have been inappropriate. In the mentioned study [29], the proportion of ST131 (21%) resistance to ciprofloxacin (54%) was almost 50% lower, and to gentamicin (52%) it was two times higher compared to our results (Table 3). Accordingly, our data suggest that the administration of ciprofloxacin and gentamicin could be estimated as being higher and lower, respectively, in our medical facilities relating to *E. coli* ESBL-producing strains compared with the previous study.

Mahaya et al. (2022) determined resistance in *E. coli* ESBL-producing strains, while resistance to carbapenems, including colistin and fosfomycin, was not detected [19]; overall, a minimal reduction in susceptibility to these agents was observed in our results (Table 3). On the other hand, strains were 100% resistant to cefotaxime and trimethoprim/sulfamethoxazole, as well as to other resistance values [19]. In comparison, in our findings, resistance was much lower, especially against amikacin and nitrofurantoin; ciprofloxacin was the only exception, against which resistance in our evaluation had a higher value (Table 3). These differences in the resistance rate may be closely related to the consumption of particular antimicrobial agents, which may be diverse between various regions and healthcare facilities, but they may also be due to the spreading of distinct multidrug-resistant bacterial clones [30].

When comparing the antimicrobial resistance of ST131 vs. non-ST131 *E. coli* ESBL-producing strains, statistically significant differences were found for ciprofloxacin (93% and 63%, respectively) in Olesen et al.'s (2013) study [21], which was consistent with our results. However, the resistance rate was slightly higher in our evaluations. Likewise, the gentamicin resistance of ST131 (16%) vs. non-ST131 (35%) [21] was also at a similar ratio in our results (Table 3). If we compare the resistance rate of ST131 strains to trimethoprim and sulfamethoxazole with non-ST131 resistance, as was performed by Olesen et al. (2013), then the resistance was higher in ST131 strains. According to this comparison, it might be inferred that ST131 could have a similar proportion of resistance to the same antimicrobial agents as other STs.

The impact of predicted risk factors related to patient characteristics and pathogen properties on antimicrobial resistance was analyzed using binary logistic regression. The male gender had the highest positive predictive value for gentamicin resistance, including ST131 subclade C2, which is associated with resistance to ciprofloxacin and ceftazidime. According to MacFadden et al.'s (2019) research comparing patient risk factors, STs, and resistance genes as predictors for antibiotic resistance in *E. coli* bloodstream infections, it was found by using a logistic regression model that ST shows strong predictive discrimination that patient epidemiologic predictors may enhance; however, the highest degree of antibiotic administration exclusion allows for the identification and presence of known resistance genes [29].

The significance of this study compared to that of others is the highlighting of the elevated emergence of ST131 in diverse samples relating to infection or colonization, as well as the predominant representation of subclade C2 instead of other sublineages, which predicted resistance to the two above-mentioned antimicrobials Moreover, we tried to find out the possible clonal transmission of ST131 in the survey, in which we were slightly successful.

The lacking in the study is that, within performed biochemical identification of *E. coli* strains, the PCR proof of these pathogens was not carried out.

Nonetheless, further examinations are required to determine the best way to make therapy more effective in monitoring and to combat antimicrobial resistance with the knowledge about the circulation of particular STs and close genetic relatedness, including the presence of their resistance determinants [31]; moreover, more studies are needed to identify when it is possible to implement tailored preventive measures to control these expanding risk clones.

4. Materials and Methods

4.1. Healthcare Facilities, Patients, and Specimens

Extended-spectrum β-lactamase-producing *Escherichia coli* strains originated from patients hospitalized at the University Hospital Bratislava (UHB) during the period from January 2017 to May 2019. The UHB is one of the largest Slovak hospitals, comprising five healthcare workplaces. The patients resided in healthcare facility No. 1, Hospital Old Town, and No. 2, Specialized Geriatric Hospital. According to our standard rules, from the patients hospitalized at the UHB, signed informed consent of the patient was obtained that his or her biological materials and samples can be provided for scientific purposes including to present the data relating to research, and such consent is in each patient's medical record if he or she was treated in our medical facility. Specimens were collected from various body sites concerning infection or colonization. Infection was specified by employing the treating physician's diagnosis, which was consistent with the clinical manifestation of disease associated with ESBL-producing *E. coli* in the particular biological sample and with the significant laboratory findings. The colonization of the patients was defined as positive cultivation when there was no evidence of infection signs at the time of the microorganisms' isolation.

4.2. Isolation, Proof of Bacterial Strains, and Susceptibility Testing

Following the primary cultivation of the samples on standard media, *Escherichia coli* strains were detected using the biochemical set ENTEROtest 24 (Erba Lachema s.r.o., Brno, Czech Republic) in our microbiology laboratory. Only one unique strain per patient was included in the study. In parallel with regular cultivation, specimens were also inoculated on COLOREX^{TM} ESBL screening chromogenic media (MkB Test a.s., Rosina, Slovak Republic). If the suspected ESBL *E. coli* strains also grew on the corresponding screening plates, they were submitted to further testing by the combination disk diffusion test with cefotaxime (30 μg) ± clavulanic acid (10 μg) and ceftazidime (30 μg) ± clavulanic acid (10 μg) for proof of ESBL production, according to the EUCAST criteria [32]. However, the Carba NP test was used to confirm carbapenemase production when testing strains showed higher breakpoint values than the 0.125 μg/mL screening cut-off for meropenem and ertapenem [32,33].

Susceptibility testing of most antimicrobial agents was performed using the colorimetric micromethod that is commercially available as a MIDITECH panel [34], and has been currently modified to test strains against the 18 antimicrobials listed in Table 3 (classes: penicillins and cephalosporins (0.25–32 μg/mL; protected by β-lactamase inhibitor, 4 μg/mL), carbapenems (ertapenem, 0.03–4 μg/mL; meropenem, 0.125–16 μg/mL), aminoglycosides (gentamicin and tobramycin, 0.125–16 μg/mL; amikacin, 0.5–64 μg/mL), tetracyclines (tetracycline, 0.125–16 μg/mL), glycylcyclines (tigecycline, 0.03–4 μg/mL), fluoroquinolones (ciprofloxacin, 0.03–4 μg/mL), polymyxins (colistin, 0.25–8 μg/mL), and antimetabolites (trimethoprim/sulfamethoxazole, 0.125/2–4/76 μg/mL)). Meanwhile, sensitivity determination against nitrofurantoin (100 μg) and fosfomycin (200 μg) was carried out with a disk diffusion assay.

4.3. DNA Isolation and NGS Library Preparation

From 15 randomly selected *E. coli* strains, nucleic acid isolation was performed using the Higher Purity^{TM} Bacterial Genomic DNA Isolation Kit (CanvaxBiotech; Valladolid, Spain). DNA concentration was measured with the Qubit ds-DNA HS Assay Kit (Invitrogen; Waltham, MA, USA). Genomic libraries for sequencing were prepared using the Nextera XT DNA Prep Kit (Illumina; San Diego, CA, USA) and purified on AMPure XP magnetic beads (Beckman Coulter Life Sciences; Brea, CA, USA). The quality of the libraries was checked by electrophoresis with the Bioanalyzer 2100 (Agilent, Santa Clara, CA, USA). Paired-end sequencing with 2 × 150 bp reads was carried out on a NextSeq system (Illumina, San Diego, CA, USA).

4.4. ST-131 Clade Multiplex PCR with fimH-30 Detection

Detection was carried out according to the work of Matsumura et al. (2017). Amplification was performed with 1× DreamTaq Green Buffer (ThermoFisher; Waltham, MA, USA); 0.2 mM dNTP (Promega; Madison, WI, USA); the primers listed in Table 4, each 0.5 pmol; 1U Taq DNA polymerase (ThermoFisher; Waltham, MA, USA); and 2 µL of template DNA, with a total volume of 25 µL. DNA was amplified using the Biometra TAdvanced (Analytik Jena; Jena, Germany) with a thermal protocol consisting of predenaturation at 98 °C for 2 min, which was followed by 30 cycles of 98 °C for 10 s, 57 °C for 20 s, and 72 °C for 40 s, and final polymerization at 72 °C for 3 min. PCR products were loaded on 2% agarose gel stained with GoodView Nucleic Acid Stain (SBS Genetech; Beijing, China) and visualized under UV light [10].

Table 4. Primers for ST131 clade multiplex PCR with *fimH-30* following Matsumura et al. (2017) [10].

Primer	Sequence	Length of Product
ST-131_F	AGCAACGATATTTGCCCATT	580 bp
ST-131_R	GGCGATAACAGTACGCCATT	
Clade C1_F	TGAATCAAAGGTCCGAGCTG	232 bp
Clade C1_R	TATGGCTGGCAGATGCTTTA	
Clade C2_F	ACGGATTCAGGTAGACGATT	164 bp
Clade C2_R	CCTCACCAAAGTTGCGATTAC	
FimH-30_F	CCGCCAATGGTACCGCTATT	354 bp
FimH-30_R	CAGCTTTAATCGCCACCCCA	

4.5. Bioinformatic Processing and Genome Analysis

Raw reads obtained from the sequencer were assembled de novo via the SPAdes assembler using standard settings (Center for Algorithmic Biotechnology; St. Petersburg State University, Saint Petersburg, Russia) [35]. Annotation of the obtained contigs was performed with the online program BV-BRC (Bacterial and Viral Bioinformatics Resource Center) and RAST (Rapid Annotation using Subsystem Technology) software. BV-BRC was also used to detect antibiotic resistance genes by implementing the NDARO (National Database of Antibiotic Resistant Organisms) database. The criterion of 100% identity and 100% coverage with the reference gene was used as the threshold for gene presence. The genomes were further analyzed using the CGE database (Center for Genomic Epidemiology; Kongens, Lyngby, Denmark) and the EnteroBase database (Warwick University). GrapeTree was used to visualize strain clusters based on the *Escherichia coli* cgMLST analysis [36].

4.6. Statistical Analysis

Considering the data related to the study's results, Pearson's chi-square ($\chi 2$) and Fisher's exact test were used alongside the statistical software IBM SPSS for Windows, version 29 (IBM SPSS Inc., Armonk, NY, USA). For the assessment of both tests, a p-value < 0.05 was statistically significant. The relevant patient- and *E. coli* strain-related factors associated with increased antimicrobial resistance were analyzed using binary logistic regression. The odds ratio and 95% confidence interval were used to quantify the estimated risk of these characteristics.

5. Conclusions

Through investigation of selected ESBL-producing *E. coli* strains from two healthcare facilities of the UHB, a three-fold higher prevalence of ST131 with a statistically significant difference was found, mainly in patients aged 65 and over, and particularly in females and from urine samples. Concerning the predominant ST131 clone, almost all types of investigated lineages and sublineages occurred the most in the first internal clinic of healthcare facility No. 1. By detecting genetic relatedness even with a small sample size of randomly selected *E. coli* strains, a possible horizontal transmission within healthcare facility No. 1 and a sporadic case between healthcare facilities No. 1 and No. 2 were identified in connection with ST131. Altogether, subclade C2 had the highest overall representation and was a significant predictor of ciprofloxacin and ceftazidime resistance; additionally, the male gender proved to be a positive predictive value for gentamicin resistance. Antimicrobial resistance was considerably higher in ST131 *E. coli* strains compared with non-ST131, with the only exception being some aminoglycosides, and indeed, these strains had a similar resistance profile to the same antimicrobial agents. Consequently, exploring these MDR worrisome and high-risk pathogenic clones is important, as well as their genetic relatedness, including recognizing genes that encode reduced susceptibility, which is crucial in the case of antimicrobial-resistant severe infections to help provide successful early therapy.

Author Contributions: Conceptualization, J.K., M.A. and H.D.; data curation, J.K., M.A., H.D., M.W. and A.L.; formal analysis, J.K., M.A., E.O. and H.D.; methodology, J.K., M.A., E.O. and H.D.; investigation, J.K., M.A., E.O. and M.W.; software, M.A. and M.W.; supervision, M.A., H.D. and A.L.; visualization, M.A. and E.O.; writing—original draft, J.K., M.A. and H.D.; writing—review and editing, M.A., H.D. and A.L.; funding acquisition, J.K., A.L. and T.M. All authors have read and agreed to the published version of the manuscript.

Funding: This publication was created thanks to the support of the Operational Program Integrated Infrastructure for the project, "Research and development in medical sciences—the way to personalize the treatment of serious neurological, cardiovascular, and cancer diseases" (code: ITMS: 313011T431), and was cofinanced by the resources of the European Regional Development Fund.

Institutional Review Board Statement: Not applicable.

Informed Consent Statement: Not applicable.

Data Availability Statement: Not applicable.

Conflicts of Interest: The authors declare no conflict of interest.

References

1. Tchesnokova, V.; Riddell, K.; Scholes, D.; Johnson, J.R.; Sokurenko, E.V. The Uropathogenic *Escherichia coli* Subclone Sequence Type 131-H30 Is Responsible for Most Antibiotic Prescription Errors at an Urgent Care Clinic. *Clin. Infect. Dis.* **2019**, *68*, 781–787. [CrossRef]
2. Gutiérrez-Gutiérrez, B.; Rodríguez-Baño, J. Current options for the treatment of infections due to extended-spectrum beta-lactamase-producing Enterobacteriaceae in different groups of patients. *Clin. Microbiol. Infect.* **2019**, *8*, 932–942. [CrossRef] [PubMed]
3. Geisinger, E.; Isberg, R.R. Interplay Between Antibiotic Resistance and Virulence During Disease Promoted by Multidrug-Resistant Bacteria. *J. Infect. Dis.* **2017**, *215*, S9–S17. [CrossRef]
4. Tamma, P.D.; Rodriguez-Baño, J. The Use of Noncarbapenem β-Lactams for the Treatment of Extended-Spectrum β-Lactamase Infections. *Clin. Infect. Dis.* **2017**, *64*, 972–980. [CrossRef]
5. Pérez-Etayo, L.; González, D.; Vitas, A.I. Clonal Complexes 23, 10, 131 and 38 as Genetic Markers of the Environmental Spread of Extended-Spectrum β-Lactamase (ESBL)-Producing *E. coli*. *Antibiotics* **2022**, *11*, 1465. [CrossRef] [PubMed]
6. Soncini, J.G.M.; Cerdeira, L.; Sano, E.; Koga, V.L.; Tizura, A.T.; Tano, Z.N.; Nakazato, G.; Kobayashi, R.K.T.; Aires, C.A.M.; Lincopan, N.; et al. Genomic insights of high-risk clones of ESBL-producing *Escherichia coli* isolated from community infections and commercial meat in southern Brazil. *Sci. Rep.* **2022**, *12*, 9354. [CrossRef] [PubMed]
7. Isendahl, J.; Giske, C.G.; Hammar, U.; Sparen, P.; Tegmark Wisell, K.; Ternhag, A.; Nauclér, P. Temporal Dynamics and Risk Factors for Bloodstream Infection With Extended-spectrum β-Lactamase-producing Bacteria in Previously-colonized Individuals: National Population-based Cohort Study. *Clin. Infect. Dis.* **2019**, *68*, 641–649. [CrossRef] [PubMed]
8. Ahmed, S.; Besser, T.E.; Call, D.R.; Weissman, S.J.; Jones, L.P.; Davis, M.A. Evaluation of two multi-locus sequence typing schemes for commensal *Escherichia coli* from dairy cattle in Washington State. *J. Microbiol. Methods* **2016**, *124*, 57–61. [CrossRef]
9. Ludden, C.; Coll, F.; Gouliouris, T.; Restif, O.; Blane, B.; Blackwell, G.A.; Kumar, N.; Naydenova, P.; Crawley, C.; Brown, N.M.; et al. Defining nosocomial transmission of *Escherichia coli* and antimicrobial resistance genes: A genomic surveillance study. *Lancet Microbe* **2021**, *2*, e472–e480. [CrossRef]
10. Matsumura, Y.; Pitout, J.D.D.; Peirano, G.; DeVinney, R.; Noguchi, T.; Yamamoto, M.; Gomi, R.; Matsuda, T.; Nakano, S.; Nagao, M.; et al. Rapid Identification of Different *Escherichia coli* Sequence Type 131 Clades. *Antimicrob. Agents Chemother.* **2017**, *61*, e00179-17. [CrossRef]
11. Mills, J.P.; Kaye, K.S.; Evans, R.; Salzman, E.; Pogue, J.; Hayakawa, K.; Marchaim, D.; Awasthy, P.; Salim, M.; Martin, E.T. Clinical and Molecular Epidemiology of Extended-Spectrum Beta-Lactamase-Producing *Escherichia Coli* Infections in Metro Detroit: Early Dominance of the ST-131 Clone. *Infect. Dis. Ther.* **2020**, *9*, 683–690. [CrossRef] [PubMed]

2. Lovayová, V.; Čurová, K.; Hrabovský, V.; Nagyová, M.; Siegfried, L.; Toporová, A.; Rimárová, K.; Andraščíková, Š. Antibiotic resistance in the invasive bacteria *Escherichia coli*. *Cent. Eur. J. Public Health* **2022**, *30*, S75–S80. [CrossRef]
3. Hrabák, J.; Empel, J.; Bergerová, T.; Fajfrlík, K.; Urbásková, P.; Kern-Zdanowicz, I.; Hryniewicz, W.; Gniadkowski, M. International clones of *Klebsiella pneumoniae* and *Escherichia coli* with extended-spectrum beta-lactamases in a Czech hospital. *J. Clin. Microbiol.* **2009**, *47*, 3353–3357. [CrossRef] [PubMed]
4. Lafolie, J.; Nicolas-Chanoine, M.H.; Grenouillet, F.; Hocquet, D.; Bertrand, X. Prevalence of *Escherichia coli* sequence type 131 and its H30 subclone among *E. coli* isolates in a French hospital. *Int. J. Antimicrob. Agents* **2014**, *44*, 466–468. [CrossRef]
5. Roer, L.; Johannesen, T.B.; Hansen, F.; Stegger, M.; Tchesnokova, V.; Sokurenko, E.; Garibay, N.; Allesøe, R.; Thomsen, M.C.F.; Lund, O.; et al. CHTyper, a Web Tool for Subtyping of Extraintestinal Pathogenic *Escherichia coli* Based on the *fumC* and *fimH* Alleles. *J. Clin. Microbiol.* **2018**, *56*, e00063-18. [CrossRef]
6. Boll, E.J.; Overballe-Petersen, S.; Hasman, H.; Roer, L.; Ng, K.; Scheutz, F.; Hammerum, A.M.; Dungu, A.; Hansen, F.; Johannesen, T.B.; et al. Emergence of Enteroaggregative *Escherichia coli* within the ST131 Lineage as a Cause of Extraintestinal Infections. *mBio* **2020**, *11*, e00353-20. [CrossRef]
7. Martischang, R.; François, P.; Cherkaoui, A.; Gaïa, N.; Renzi, G.; Agostinho, A.; Perez, M.; Graf. C.E.; Harbarth, S. Epidemiology of ESBL-producing *Escherichia coli* from repeated prevalence studies over 11 years in a long-term-care facility. *Antimicrob. Resist. Infect. Control* **2021**, *10*, 148. [CrossRef] [PubMed]
8. Lindblom, A.; Kiszakiewicz, C.; Kristiansson, E.; Yazdanshenas, S.; Kamenska, N.; Karami, N.; Åhrén, C. The impact of the ST131 clone on recurrent ESBL-producing *E. coli* urinary tract infection: A prospective comparative study. *Sci. Rep.* **2022**, *12*, 10048. [CrossRef]
9. Mhaya, A.; Trabelsi, R.; Aillerie, S.; M'Zali, F.; Bégu, D.; Tounsi, S.; Gdoura, R.; Arpin, C. Detection of Clones B2-ST131-C2 and A-ST617 in *Escherichia coli* Producing Both CTX-M-15 and CTX-M-27 from Tunisian Community Patients. *Antibiotics* **2022**, *11*, 1329. [CrossRef]
0. Ellaby, N.; Doumith, M.; Hopkins, K.L.; Woodford, N.; Ellington, M.J. Emergence of diversity in carbapenemase-producing *Escherichia coli* ST131, England, January 2014 to June 2016. *Eurosurveillance* **2019**, *24*, 1800627. [CrossRef]
1. Olesen, B.; Hansen, D.S.; Nilsson, F.; Frimodt-Møller, J.; Leihof, R.F.; Struve, C.; Scheutz, F.; Johnston, B.; Krogfelt, K.A.; Johnson, J.R. Prevalence and characteristics of the epidemic multiresistant *Escherichia coli* ST131 clonal group among extended-spectrum beta-lactamase-producing *E. coli* isolates in Copenhagen, Denmark. *J. Clin. Microbiol.* **2013**, *51*, 1779–1785. [CrossRef] [PubMed]
2. Weterings, V.; van den Bijllaardt, W.; Bootsma, M.; Hendriks, Y.; Kilsdonk, L.; Mulders, A.; Kluytmans, J. Duration of rectal colonization with extended-spectrum beta-lactamase-producing *Escherichia coli*: Results of an open, dynamic cohort study in Dutch nursing home residents (2013–2019). *Antimicrob. Resist. Infect. Control* **2022**, *11*, 98. [CrossRef]
3. Feng, Y.C.; Liou, C.H.; Ng, W.V.; Chen, F.J.; Hung, C.H.; Liu, P.Y.; Liao, Y.C.; Wu, H.C.; Cheng, M.F. Distribution and Genomic Characterization of Third-Generation Cephalosporin-Resistant *Escherichia coli* Isolated from a Single Family and Home Environment: A 2-Year Longitudinal Study. *Antibiotics* **2022**, *11*, 1152. [CrossRef] [PubMed]
4. van den Bunt, G.; Fluit, A.C.; Bootsma, M.C.J.; van Duijkeren, E.; Scharringa, J.; van Pelt, W.; Bonten, M.J.M. Dynamics of Intestinal Carriage of Extended-Spectrum Beta-lactamase-Producing Enterobacteriaceae in the Dutch General Population, 2014–2016. *Clin. Infect. Dis.* **2020**, *71*, 1847–1855. [CrossRef] [PubMed]
5. Flament-Simon, S.C.; García, V.; Duprilot, M.; Mayer, N.; Alonso, M.P.; García-Meniño, I.; Blanco, J.E.; Blanco, M.; Nicolas-Chanoine, M.H.; Blanco, J. High Prevalence of ST131 Subclades C2-*H*30Rx and C1-M27 among Extended-Spectrum β-Lactamase-Producing *Escherichia coli* Causing Human Extraintestinal Infections in Patients from Two Hospitals of Spain and France during 2015. *Front. Cell. Infect. Microbiol.* **2020**, *10*, 125. [CrossRef] [PubMed]
6. Tchesnokova, V.L.; Rechkina, E.; Larson, L.; Ferrier, K.; Weaver, J.L.; Schroeder, D.W.; She, R.; Butler-Wu, S.M.; Aguero-Rosenfeld, M.E.; Zerr, D.; et al. Rapid and Extensive Expansion in the United States of a New Multidrug-resistant *Escherichia coli* Clonal Group, Sequence Type 1193. *Clin. Infect. Dis.* **2019**, *68*, 334–337. [CrossRef] [PubMed]
7. Roberts, L.W.; Hoi, L.T.; Khokhar, F.A.; Hoa, N.T.; Giang, T.V.; Bui, C.; Ninh, T.H.; Co, D.X.; Binh, N.G.; Long, H.B.; et al. Genomic characterisation of multidrug-resistant *Escherichia coli, Klebsiella pneumoniae*, and *Acinetobacter baumannii* in two intensive care units in Hanoi, Viet Nam: A prospective observational cohort study. *Lancet Microbe* **2022**, *3*, e857–e866. [CrossRef]
8. Sadek, M.; Saad, A.M.; Nordmann, P.; Poirel, L. Genomic Characterization of an Extensively Drug-Resistant Extra-Intestinal Pathogenic (ExPEC) *Escherichia coli* Clinical Isolate Co-Producing Two Carbapenemases and a 16S rRNA Methylase. *Antibiotics* **2022**, *11*, 1479. [CrossRef]
9. MacFadden, D.R.; Melano, R.G.; Coburn, B.; Tijet, N.; Hanage, W.P.; Daneman, N. Comparing Patient Risk Factor-, Sequence Type-, and Resistance Locus Identification-Based Approaches for Predicting Antibiotic Resistance in *Escherichia coli* Bloodstream Infections. *J. Clin. Microbiol.* **2019**, *57*, e01780–e18. [CrossRef]
0. Cummins, E.A.; Snaith, A.E.; McNally, A.; Hall, R.J. The role of potentiating mutations in the evolution of pandemic *Escherichia coli* clones. *Eur. J. Clin. Microbiol. Infect. Dis.* **2021**, *40*, 1–10. [CrossRef]
1. van Hout, D.; Verschuuren, T.D.; Bruijning-Verhagen, P.C.J.; Bosch, T.; Schürch, A.C.; Willems, R.J.L.; Bonten, M.J.M.; Kluytmans, J.A.J.W. Extended-spectrum beta-lactamase (ESBL)-producing and non-ESBL-producing *Escherichia coli* isolates causing bacteremia in the Netherlands (2014–2016) differ in clonal distribution, antimicrobial resistance gene and virulence gene content. *PLoS ONE* **2020**, *15*, e0227604. [CrossRef] [PubMed]

32. EUCAST. Breakpoint Tables for Interpretation of MICs and Zone Diameters. Version 13.0. 2023. Available online: https: www.eucast.org/fileadmin/src/media/PDFs/EUCAST_files/Breakpoint_tables/v_13.0_Breakpoint_Tables.pdf (accessed on January 2023).
33. Nordmann, P.; Poirel, L.; Dortet, L. Rapid Detection of Carbapenemase-producing Enterobacteriaceae. *Emerg. Infect. Dis.* **2012**, 1503–1507. [CrossRef] [PubMed]
34. Gattringer, R.; Nikš, M.; Ostertág, R.; Schwarz, K.; Medvedovic, H.; Graninger, W.; Georgopoulos, A. Evaluation of MIDITEC automated colorimeter MIC reading for antimicrobial susceptibility testing. *J. Antimicrob. Chemother.* **2002**, *49*, 651–659. [CrossRe [PubMed]
35. Prjibelski, A.D.; Antipov, D.; Meleshko, D.; Lapidus, A.L.; Korobeynikov, A.I. Using SPAdes De Novo Assembler. *Curr. Prot Bioinform.* **2020**, *70*, e102. [CrossRef]
36. Zhou, Z.; Alikhan, N.F.; Mohamed, K.; Fan, Y.; Agama Study Group; Achtman, M. The EnteroBase user's guide, with case studi on *Salmonella* transmissions, *Yersinia pestis* phylogeny, and *Escherichia* core genomic diversity. *Genome Res.* **2020**, *30*, 138–1 [CrossRef] [PubMed]

antibiotics

Article

Moving toward Extensively Drug-Resistant: Four-Year Antimicrobial Resistance Trends of *Acinetobacter baumannii* from the Largest Department of Internal Medicine in Slovakia

Yashar Jalali [1], Adriána Liptáková [2], Monika Jalali [1] and Juraj Payer [1,*]

[1] Faculty of Medicine, Comenius University in Bratislava, 5th Department of Internal Medicine, University Hospital Bratislava, Ružinov, Špitálska 24, 813 72, and Ružinovská 4810/6, 821 01 Bratislava, Slovakia; yashar.jalali@fmed.uniba.sk (Y.J.); adamcova.monika@gmail.com (M.J.)
[2] Institute of Microbiology, Faculty of Medicine, Comenius University in Bratislava, Špitálska 24, 813 72 Bratislava, Slovakia; adriana.liptakova@fmed.uniba.sk
* Correspondence: payer@ru.unb.sk

Abstract: *A. baumannii* imposes a great burden on medical systems worldwide. Surveillance of trends of antibiotic resistance provides a great deal of information needed for antimicrobial stewardship programmes nationwide. Clinical data from long-term, continuous surveillance on trends of antibiotic resistance of *A. baumannii* in Slovakia is missing. One hundred and forty-nine samples of *A. baumannii* were isolated over a period of four years. A panel of 19 antibiotics from seven antibiotic categories were tested for the bacterium's susceptibility. Resistance results were evaluated, and the significance of patterns was estimated using simple linear regression analysis. All isolates were more than 85% resistant to at least 13 out of the 19 tested antibiotics. A significant rise in resistance was recorded for aminoglycosides and imipenem from 2019 to 2022. Colistin and ampicillin-sulbactam have been the only antibiotics maintaining more than 80% efficacy on the bacterium to date. A significant rise in extensively drug-resistant (XDR) strains among carbapenem-resistant (CR) isolates has been recorded. Multidrug-resistance (MDR) among all *A. baumannii* isolates and XDR among CR strains of the bacterium have risen significantly in the last four years.

Keywords: *Acinetobacter baumannii*; antibiotic resistance trends; multidrug-resistant; carbapenem-resistant; healthcare-associated infection

1. Introduction

Acinetobacter is a ubiquitous opportunistic coccobacillus, having reservoirs in nearly all environmental niches [1–3]. The natural habitats of *Acinetobacter baumannii* are still poorly understood, as it is almost exclusively isolated in close-contact communities and hospital environments [3,4]. Despite being discovered in the early years of the previous century, the bacterium has begun to attract international attention only recently due to a significant rise in isolation of its MDR and even pan-drug-resistant (PDR) strains worldwide [4–6]. The bacterium's strong ability to resist the effects of a wide range of antibiotics, due to both intrinsic and acquired mechanisms, has greatly increased the morbidities and the mortalities caused by *A. baumannii* in recent years [1,4,6,7]. Combined resistance (resistance to fluoroquinolones, aminoglycosides and carbapenems) is a term utilised by the European Centre for Disease Prevention and Control (ECDC) to present epidemiology and trends of drug resistance changes of *Acinetobacter* species [8]. *Acinetobacter* species resistant to one antibiotic in all three tested combined antibiotic families can be regarded as MDR species. The percentage of combined resistance isolates of *Acinetobacter* species, in the last published surveillance by the ECDC in 2021, varies widely in Europe, from 7% to above 50% (most intensive in southern and eastern European countries) [8]. The rise in the percentage of combined resistant isolates of the *Acinetobacter* species in Slovakia from 24% in 2020 to 53% in 2021 is alarming [8].

Resistance to carbapenem alone is considered to be a marker of extensive antimicrobial resistance, since it involves a large range of co-resistance to other unrelated antibiotic classes [6]. Six years after the listing of CR *A. baumannii* by the World Health Organisation (WHO) as a critical priority bacterium for which new antibiotics are urgently needed, no real development has occurred [1,4,7,9]. The rate of clinical outbreaks of CR *A. baumannii* ranges between 1% and over 30% in Europe, being again most intense in eastern and south-eastern countries [1].

Slovakia's location in a region with some of highest percentages of combined resistant isolates of *Acinetobacter* species (Hungary: 80%, Romania: 89%, Poland: 67%, Czechia: 50%, Croatia: 98%) makes the surveillance of trends of the bacterium's antibiotic resistance of great importance [8]. This data provides insight into the expected prevalence and trends of the antibiotic resistance of *A. baumannii*, which is needed to develop the rationale for antibiotic stewardship policies battling against the bacterium not only in Slovakia but the whole region.

This study presents comprehensive data from four years of surveillance (2019 to 2022) on trends of antibiotic resistance of all clinically relevant *A. baumannii* isolates from the largest department of internal medicine in Slovakia.

2. Results

2.1. Patients Characteristics and Demographic Data

A total of 11,856 patients were hospitalised during the study period in our department (3618 patients in 2019, 2940 patients in 2020, 2579 patients in 2021, and 2719 patients in 2022). A total of 149 samples of *A. baumannii* were isolated. Four samples were isolated from patients hospitalised in intensive care unit (ICU) (4/149, 3%), 13 samples were isolated from patients hospitalised in intermediary ward for critically ill patients (13/149, 9%) and 132 samples were isolated from patients hospitalised in standard male and female wards (132/149, 88%) during study period. Average hospitalisation in ICU was recorded seven to ten days, and in other wards 25 days. Male patients were more likely to be infected by *A. baumannii* than female patients (57% to 43%). The mean age of male patients was 12 years higher than that of female patients (Table 1).

Table 1. Number of isolated *A. baumannii*, prevalence, multidrug-resistance, and extensively drug-resistant traits.

	Total	2019	2020	2021	2022
Number of hospitalised patients	11,856	3618	2940	2579	2719
Number of isolated *A. baumannii*	149	28	28	52	41
Number of male patients (n/N, cases %)	85/149, 57%				
Age of male patients (Mean)	72.7 (SD ± 18.2)				
Number of female patients (n/N, cases %)	64/149, 43%				
Age of female patients (Mean)	60.2 (SD ± 20.2)				
Hospitalisation duration (Av. days)	25.3				
Healthcare-associated infection (n/N) Cases (%)	126/149 85%	17/28 60%	24/28 85%	48/52 92%	37/41 90%
Associated mortality rate (n/N) Cases (%)	61/149 40%	11/28 39%	12/28 43%	23/52 44%	15/41 37%
Prevalence (P^1 %)		0.8%	1%	2%	1.5%
Number of MDR strains among all isolates (n/N) Cases (%)	78/149 52%	9/28 32%	12/28 43%	35/52 67%	22/41 53%
Number of XDR strains among all isolates (n/N) Cases (%)	50/149 33%	9/28 32%	12/28 42%	14/52 26%	15/41 36%

Av.: average, N: total number of isolated samples, n: number of isolated samples for given variable (for example: healthcare-associated infection (n/N) means number of isolated healthcare-associated infection samples to total number of isolated samples, also presented in percent form), P^1: true prevalence. SD: standard deviation, XDR: extensively drug-resistant.

The prevalence of *A. baumannii* was 0.8% in 2019, 1% in 2020, 2% in 2021, and 1.5% in 2022. MDR trends changed from 32% in 2019 to 43% in 2020 due to an increase in resistance to tobramycin among isolated strains. This trend further changed to 67% in 2021 and 53% in 2022 (Table 1). There was a significant association between the trend of prevalence of the bacterium and its MDR (R^2 = 0.97, p = 0.01 confidence level (CL): 95%). The total healthcare-associated infection (HAI) rate was 85% over the four years. We recorded a steady rise in the percentage of HAI, from 60% in 2019 to 90% in 2022. Despite similarities in patterns of HAI and prevalence of the bacterium (Table 1), there was no statistically significant association between them (R^2 = 0.61, p = 0.21 CL: 95%). The associated mortality rate (AMR), defined as the effect of the infection on patient death as an outcome, in association with other comorbidities, is shown in Table 1. The AMR percentage (number of diagnosed patients with *A. baumannii* who died during the same hospitalisation without laboratory data suggestive of infection management divided by the number of all diagnosed cases of *A. baumannii*, expressed as a percentage) was 40% in total, meaning 40% of patients diagnosed with the bacterium died during the same hospitalisation. There were no associations between AMR and the MDR trend, prevalence, and/or HAI.

2.2. Distribution of Samples' Isolation

A. baumannii was predominantly isolated from respiratory secretions (n: 64/149, 43%), followed by wounds or decubitus samples (n: 44/149, 30%), urinary tract infections (n: 32/149, 21%), haemocultures, and samples from central venous catheter-related infections (n: 9/149, 6%) (Figure 1).

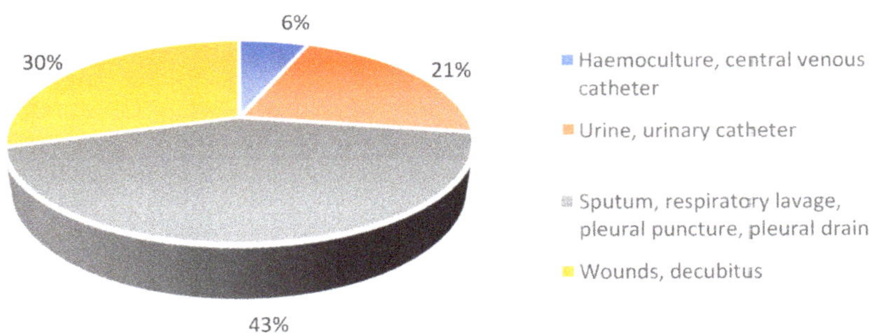

Figure 1. Frequency of isolation (%) of *A. baumannii* from a given site.

2.3. Demographic Data and Prevalence of Carbapenem Resistant Isolates

The number of CR strains among all isolates, despite an initial rise during the years 2019 and 2020 (68% to 75%, respectively), declined to 39% and 44% in 2021 and 2022, respectively. The percentage of multidrug-resistant cases hovered steadily below 25% among all CR isolates. However, alarmingly, the number of extensively drug-resistant strains rose continuously among CR isolates, reaching 83% of cases in the year 2022 (Table 2).

We recorded an increase in the rate of HAI from CR strains during the study period. The association between the increase in percentage of HAI and the percentage of XDR isolation was statistically significant (R^2 = 0.97, p = 0.01, CL: 95%), meaning the higher the percentage of HAI, the higher the probability of isolation of the XDR bacterium among CR strains.

Table 2. Carbapenem-resistant strains of *A. baumannii*, their prevalence, and resistant traits.

	Total	2019	2020	2021	2022
Number of CR strains among all isolates (n/N) Cases (%)	78/149 52%	19/28 68%	21/28 75%	20/52 39%	18/41 44%
Prevalence of CR isolates (P^2 %)		0.5%	0.7%	0.7%	0.6%
Healthcare-associated infection among CR isolates (n/N) Cases (%)	62/78 79%	12/19 63%	16/21 76%	17/20 85%	17/18 94%
Associated mortality rate among CR isolates (n/N) Cases (%)	36/78 46%	9/19 50%	10/21 48%	10/20 50%	7/18 39%
Number of MDR cases among CR isolates (n/N) Cases (%)	15/78 19%	4/19 21%	3/21 14%	5/20 25%	3/18 16%
Number of XDR strains among CR isolates (n/N) Cases (%)	50/78 64%	9/19 47%	12/21 57%	14/20 70%	15/18 83%

CR: carbapenem-resistant, MDR: multidrug-resistant, N: total number of isolated samples, n: number of isolated samples for given variable (for example: healthcare-associated infection (n/N) means number of isolated healthcare-associated infection samples to total number of isolated samples, also presented in percent form), P^2: carbapenem-resistance prevalence, XDR: extensively drug-resistant.

2.4. Trends of Antibiotic Resistance

A susceptibility test was performed on the panel of 19 antibiotics from seven antibiotic families as the routine antibiogram examination on the bacterium. The antibiotic families included penicillins (ampicillin, ampicillin-sulbactam, and piperacillin-tazobactam), cephalosporines (cefotaxime, ceftazidime, cefuroxime, cefepime, and cefoperazone-sulbactam), carbapenems (imipenem, ertapenem, and meropenem), monobactams (aztreonam), fluroquinolones (ciprofloxacin), aminoglycosides (amikacin, gentamicin, and tobramycin), tetracyclines (tetracycline), sulphonamides (trimethoprim-sulfamethoxazole), and polymyxins (colistin).

All isolated strains of *A. baumannii* showed 100% resistance to ertapenem, aztreonam, tetracycline, and cefuroxime. Resistance to ampicillin, cefotaxime, and cefoperazone-sulbactam was steadily above 90% (Figure 2 and Table 3). Resistance to aminoglycosides (amikacin and tobramycin) showed a significant rise from 11% in 2019 to 85% in 2022 for amikacin ($R^2 = 0.97$, $p = 0.01$) and from 9% in 2019 to 85% in 2022 for tobramycin ($R^2 = 0.96$, $p = 0.01$). Although resistance to gentamycin doubled (from 21% in 2019 to 54% in 2022) in the period of four years (Figure 2 and Table 3), it was not a statistically significant increase ($R^2 = 0.48$, $p = 0.30$). The increase in antibiotic resistance continued for all tested antibiotics in the penicillin, cephalosporine, and carbapenem families with the exception of ampicillin-sulbactam. The same increase in resistance was recorded in all isolated cases for ciprofloxacin (from 54% in 2019 to 90% in 2022, $R^2 = 0.68$, $p = 0.17$) and trimethoprim-sulfamethoxazole (from 50% in 2019 to 88% in 2022, $R^2 = 0.72$, $p = 0.14$). However, none of the trends of resistance in the tested antibiotics (except for imipenem, amikacin, and tobramycin), despite a positive linear projection plotted on the graph, was statistically significant (Table 3, Figure 2). Data regarding antibiotic resistance trends for separate antibiotic categories are provided in Supplementary File S1.

Table 3. Antibiotic resistance trends of all isolated *A. baumannii* from 2019 to 2022.

	2019	2020	2021	2022	R^2-p (CL: 95%)
ATB	No. R/S (%)	No. R/S (%)	No. R/S (%)	No. R/S (%)	
Ampicillin	28/0 (100%)	24/4 (86%)	51/1 (98%)	41/0 (100%)	-
Ampicillin-sulbactam	1/27 (4%)	9/23 (32%)	14/38 (27%)	6/34 (15%)	-
Piperacillin-tazobactam	15/13 (54%)	24/4 (86%)	47/5 (94%)	37/4 (90%)	0.67–0.18

Table 3. *Cont.*

	2019	2020	2021	2022	R^2-p (CL: 95%)
Cefotaxime	28/0 (100%)	28/0 (100%)	48/4 (92%)	37/4 (90%)	-
Ceftazidime	15/13 (54%)	24/4 (86%)	48/4 (92%)	37/4 (90%)	0.68–0.17
Cefuroxime	28/0 (100%)	28/0 (100%)	52/0 (100%)	41/0 (100%)	-
Cefepime	14/14 (50%)	20/8 (71%)	49/3 (94%)	35/6 (85%)	0.74–0.13
Cefoperazone-sulbactam	28/0 (100%)	28/0 (100%)	51/1 (98%)	41/0 (100%)	-
Imipenem	18/10 (64%)	23/5 (82%)	45/7 (87%)	40/1 (98%)	0.94–0.02
Meropenem	14/14 (50%)	24/4 (86%)	46/6 (88%)	37/4 (90%)	0.68–0.17
Ertapenem	28/0 (100%)	28/0 (100%)	52/0 (100%)	41/0 (100%)	-
Aztreonam	28/0 (100%)	28/0 (100%)	52/0 (100%)	41/0 (100%)	-
Ciprofloxacin	15/13 (54%)	24/4 (86%)	48/4 (92%)	37/4 (90%)	0.68–0.17
Amikacin	3/15 (11%)	13/15 (46%)	34/18 (65%)	356 (85%)	0.97–0.01
Tobramycin	3/25 (9%)	13/15 (46%)	35/17 (67%)	35/6 (85%)	0.96–0.01
Gentamycin	6/22 (21%)	15/13 (54%)	3913 (75%)	22/19 (54%)	-
Tetracycline	28/0 (100%)	28/0 (100%)	52/0 (100%)	41/0 (100%)	-
Trimethoprim sulfamethoxazole	14/14 (50%)	23/5 (82%)	45/7 (87%)	36/5 (88%)	0.72–0.14
Colistin	0/28 (0%)	0/28 (0%)	2/49 (4%)	0/41 (0%)	-

ATB: antibiotic, No: number, R: resistance, S: susceptibility, R^2: R square (coefficient of determination), CL: confidence level.

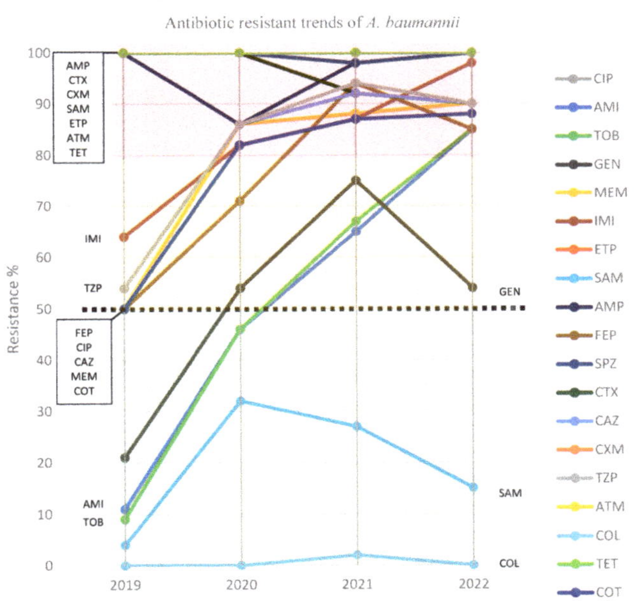

Figure 2. Antibiotic resistance trends of *A. baumannii*. AMI: amikacin, AMP: ampicillin, ATM: aztreonam, CAZ: ceftazidime, CIP: ciprofloxacin, COL: colistin, COT: co-trimoxazole (trimethoprim-sulfamethoxazole), CTX: cefotaxime, CXM: cefuroxime, ETP: ertapenem, FEP: cefepime, GEN: gentamycin, IMI: imipenem, MEM: meropenem, SAM: ampicillin-sulbactam, SPZ: sulperazone (cefoperazone-sulbactam), TET: tetracycline, TOB: tobramycin, TZP: tazocin (piperacillin-tazobactam).

The significance of resistance trends for antibiotics with horizontal projection observed on the graph was not calculated.

2.5. Trends of Antibiotic Resistance among CR Isolates

Four years of antibiotic resistance trends for isolated CR strains of *A. baumannii* are shown in Figure 3. Four columns are displayed above each tested antibiotic, with each representing a year from 2019 to 2022. Accumulated percentage of susceptibility (in blue) is shown in contrast to accumulated percentage of resistance (in orange) to a given antibiotic in a given year for each column.

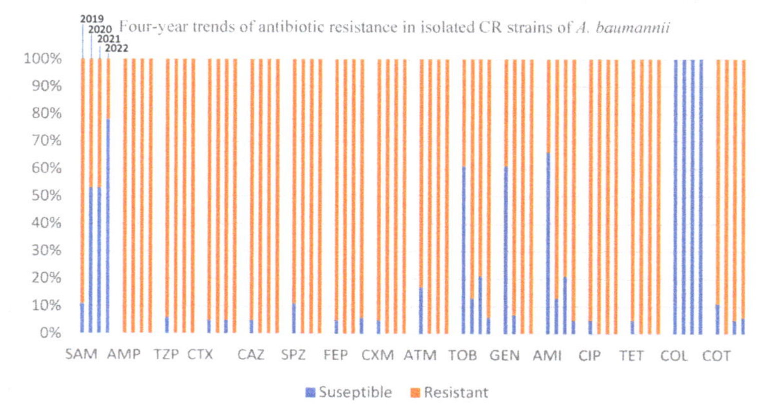

Figure 3. Four-year trends of antibiotic resistance in isolated CR strains of *A. baumannii*. AMI: amikacin, AMP: ampicillin, ATM: aztreonam, CAZ: ceftazidime, CIP: ciprofloxacin, COL: colistin, COT: co-trimoxazole (trimethoprim-sulfamethoxazole), CTX: cefotaxime, CXM: cefuroxime, FEP: cefepime, GEN: gentamycin, SAM: ampicillin-sulbactam, SPZ: sulperazone (cefoperazone-sulbactam), TET: tetracycline, TOB: tobramycin, TZP: tazocin (piperacillin-tazobactam).

All isolated CR strains of the bacterium showed 100% resistance to ampicillin, piperacillin-tazobactam, cefotaxime, ceftazidime, cefoperazone-sulbactam, cefuroxime, aztreonam, gentamycin, ciprofloxacin, and tetracycline in 2022 (Figure 3). Susceptibility to the rest of the tested antibiotics (except for ampicillin-sulbactam and colistin) was less than 10% (Figure 3), reflecting the 83% XDR phenotype of all CR isolates of the bacterium in this year. Resistance to carbapenems (as expected) was strongly associated with the XDR phenotype (64% vs. 3%, $p = 0.001$).

3. Discussion

This study provides the first clinical data from a long-term continuous surveillance programme on epidemiology, HAI, associated mortality rate, and trends in antibiotic resistance of *A. baumannii* from Slovakia.

A high percentage of HAI was recorded in our study. Nearly 85% of all samples were isolates from patients who acquired the bacterium during their current hospitalisation. The association between duration of hospitalisation and increased risk of acquiring HAI has been demonstrated in several publications [10–12]. It is fairly widely accepted that the longer the patient is hospitalised, the higher their chance of contracting a HAI. Hence, minimum average duration of hospitalisation until diagnosis of HAI with a given bacterium is of great epidemiological and clinical importance. In our previous published report on the association between HAI *A. baumannii* infection and duration of hospitalisation (from July 2019 to October 2020), we demonstrated that an average of 14 days' hospitalisation was associated with elevated risk of being infected with strains of the bacterium in our department [13]. This trend was similar both in MDR and CR isolates [13]. In the current study, however, patients were hospitalised for 25 days on average before being diagnosed with *A. baumannii*-related infections. The difference between these two data could possibly be explained by the larger cohort of patients in the current study. Since data from epidemiological studies on the state of HAIs of *A. baumannii* from a nationwide database or other single-centre studies in Slovakia are missing, optimisation of this average based on a comparison of available data is not possible for the authors. However, similar to our data, an average of 15 days (or longer) of hospitalisation prior to isolation of *A. baumannii* has been demonstrated in studies from other countries [14–16].

In several studies, the presence of HAI has been shown to significantly contribute to all-cause mortality with a non-significant odds ratio [16–19]. These data suggest that the higher mortality rate in HAI isolates is associated with the higher prevalence of virulent phenotypes due to increased antimicrobial treatment exposure [16]. Alrahmany et al., demonstrated in a recent study that prolonged hospitalisation of more than seven days was a significant predictor of higher mortality rate in *A. baumannii*-related infections ($p < 0.05$) [16]. They argued that 65% of patients who were diagnosed with MDR *A. baumannii* infection during hospitalisation were hospitalised on average for more than seven days, significantly contributing to the outcome of the disease. These data correlate with our findings, demonstrating a significant association between HAI and isolation of XDR strains ($R^2 = 0.97$, $p = 0.01$, CL: 95%). As prolonged hospitalisation is an important contributor to the increased cost associated with infection complications, efforts to minimise the nosocomial transmission of HAI should be encouraged. Therefore, such HAI-reducing efforts, along with the resultant decreased mortality and cost, also provide the benefits of reduction in risk of transmission of other MDR organisms and the use of broad-spectrum antibiotics.

In addition to prolonged hospitalisation and HAI, several other risk factors have been identified for the acquisition of antibiotic-resistant *A. baumannii*. Some of these risk factors include prior colonisation with the bacterium, invasive procedures, and ICU stay [20–22].

Since only 4 of 149 (3%) of samples were isolated from patients hospitalised in the ICU, an association between ICU hospitalisation and acquisition of *A. baumannii*, unlike the mentioned reports [23–25], was not demonstrated in our study. Lack of such a relationship could be explained by the relatively low percentage (34%) of bacterium isolation associated with prior invasive procedures (in total), and the short average duration of ICU hospitalisation in the department (average of seven to ten days).

Urinary catheterisation (UC) along with central venous catheterisation (CVC) were the two most common invasive procures associated with acquisition of *A. baumannii* infection in our study (nearly 34% of cases). All patients from whom the bacterium was isolated from urinary tract infections (UTI) also had UC at the time of sampling (30% of total isolated cases). Hence, UC was the most important invasive procedure considered a risk factor for acquisition of *A. baumannii* infection. Six out of nine isolated blood-borne *A. baumannii* infections were associated with CVC (4% of total isolated cases). Although the highest number of samples (43%) were isolated from respiratory tract-related infections, no invasive ventilation was recorded for any of the patients.

Due to the defined exclusion criteria of the study, an evaluation of association of prior colonisation with the bacterium was not performed.

Old age is generally associated with higher numbers of comorbidities, immunocompromised status, prolongation of hospitalisation, and increased disease severity. The predominance of infection by *A. baumannii* bacteria in males (57% compared to 43% in females, Table 1), in this sense, could be partly explained by the age difference between the male and female patients (mean of 12 years), which included an older cohort of male patients.

Several statistically significant changes in antibiotic resistance rates of *A. baumannii* were observed between January 2019 and December 2022. All isolates of *A. baumannii* were nearly 100% resistant to tetracycline, aztreonam, ertapenem, cefoperazone-sulbactam, and cefuroxime throughout the study period. By the end of 2022, more than 85% resistance was recorded in the other 11 tested antibiotics. Although statistically, the model of resistance pattern of these antibiotics with positive linear trajectory did not fit the confidence level, the percent of susceptibility of the bacterium was recorded at less than 15% for all of them. A significant rise in the pattern of resistance was recorded for amikacin ($R^2 = 0.97$, $p = 0.01$), tobramycin ($R^2 = 0.96$, $p = 0.01$), and imipenem (R^2 [26] $= 0.94$, $p = 0.02$). An exception to this rising resistance pattern was recorded only for colistin (0% resistance), ampicillin-sulbactam (15% resistance), and gentamycin (54% resistance), which are known as last-resort antibiotics left as treatment options for the bacterium [27,28].

An annual growth of 4% in broad-spectrum antimicrobial consumption in 2021 alone in Slovakia (surveillance report of antimicrobial consumption in EU/EEA issued by ECDC) places the country among the worst in Europe (just after Poland, Greece, and Croatia) [29]. Demonstrating the extent and rate of resistance changes of *A. baumannii* in light of broad-spectrum antibiotic consumption growth is worrisome, emphasizing the urgent need for a stricter nationwide antimicrobial stewardship programme.

To date, the only epidemiological data regarding the antibiotic resistance of *A. baumannii* in Slovakia can be acquired from a surveillance atlas of infectious disease published by the ECDC [8]. These data, however, have their limitations. Firstly, the estimate of national population coverage (for Slovakia in the last published report, this was 56%) does not include *A. baumannii* due to outliers in some countries [30]. Hence, the estimation of the epidemiological significance of the presented percentage for smaller or larger samples is unclear. Secondly, it provides information on the resistance of the bacterium to only three antibiotic families (aminoglycosides, fluroquinolones, and carbapenems). As our data demonstrated, susceptibility of the bacterium to ampicillin-sulbactam, for example, not only remained low but showed a decrease in resistance trend in last two years (Figure 2). Such clinically important information cannot be obtained from the ECDC report.

Our data demonstrates that the MDR of *A. baumannii* nearly doubled in the last four years in our department. Importantly, the percentage of MDR isolates in our cohort of patients was higher both in 2020 and 2021 in comparison with national data obtained from the surveillance atlas of infectious disease published by the ECDC (data from 2022 have not yet been published) [8]. This difference could likely be due to inappropriate antibiotic utilisation or cross-acquisition of resistance in our department. There was a significant association between patterns of MDR and prevalence of the bacterium ($R^2 = 0.97$, $p = 0.01$ confidence level (CL): 95%). This means that an increase of antibiotic resistance in the bacterium could predict an increase in its prevalence in the future. In a recent study on the prevalence of *A. baumannii* published by Chaoying Ma et al. [1], using data from the European Antimicrobial Resistance Surveillance Network (EARS-Net) and the Central Asian and European Surveillance of Antimicrobial Resistance Network (CAESAR) from the WHO, it was demonstrated that the relation between prevalence and resistance can be explained by the exponential regression model. The authors stated that once antibiotic resistance rates are low, the isolation rate of the bacterium is expected to be stable, but once the resistance rate increases by 50% or more, a surge in prevalence could be expected. Although we did not record the expected surge in the percentage of prevalence in our study (perhaps due to the monocentric nature of the study and small sample size), a clear association between the two variables was demonstrated.

Because of its great importance in clinical practice (due to involvement of large range of co-resistance to other unrelated antibiotic classes), CR *A. baumannii* has been the centre of attention in research on antibiotic resistant trends of the bacterium in recent years. Due to this importance, we performed separate analyses on the antibiotic resistance data of CR isolates of *A. baumannii* in our study. Although numbers of CR isolates were fairly stable (on average 52% of cases over the four years) and prevalence of CR strains stayed below 1%, the percentage of XDR strains among CR isolates rose significantly nonetheless, showing a significant association between CR and XDR phenotypes (64% vs. 3%, $p = 0.001$). Since resistance to carbapenems is considered to be a marker of extensive antimicrobial resistance [31–33], this finding was not surprising. However, the speed of development of the XDR phenotype (from 47% in 2019 to 83% in 2022) among CR isolates is worrisome. It is noteworthy to mention that all isolated XDR strains were CR. The pattern of resistance over four years changed in such a way that the bacterium's susceptibility to amikacin and tobramycin dropped to less than 10% and susceptibility to eight other antibiotics dropped to 0% (Figure 3), marking the rapid movement of CR strains toward being completely XDR in the coming years. In a recent study published by Appaneal et al., on a cohort of 4599 patients diagnosed with *A. baumannii* infection from a nationwide database in the USA [34], diagnosis with *A. baumannii* was stated as an indicator of poor outcome in total for all patients. Moreover, they explained that poor outcome was significantly associated with MDR and/or CR infection phenotypes of the bacterium. The associated mortality rate of the bacterium in our study stayed at around 40%, in accordance with data published by Appaneal et al., However, although slightly higher, the associated mortality rates of CR and non-CR isolates were not significantly different (40% vs. 46%).

Since genome-based strain typing and whole-genome sequencing data on the antibiotic resistance of *A. baumannii* strains in Slovakia is still unavailable, we may only estimate the diversity of the gene pool the bacteria might have. Several studies have demonstrated horizontal gene transfer of mobile antimicrobial-resistant genetic elements (known as mobile gene element MEG) to and from *A. baumannii* through plasmids, integrons, insertion sequences, transposons, prophages, etc. [35–38]. *K. pneumoniae* is one of the most well-studied bacteria, known as a main pool of drug resistance and as a transporter of resistance genes among Gram-negative bacteria, including *A. baumannii* [37,39,40]. In a recent study published by Koreň et al., a whole-genome sequencing of CR strains of *K. pneumoniae*, isolated from the Bratislava University Hospital complex, was performed [41,42]. The results indicated a high presence of NDM-1, SHV-11, SHV-168, CTX-M-15, OXA-1, and KPC-2-encoded serine carbapenemase genes responsible for the carbapenemase activity of isolated strains. That could only serve as an indication of the antibiotic-resistant genetic pool and MGEs of circulating *A. baumannii* strains in the same healthcare facility.

Limited available antimicrobial treatment along with growing resistance to scarce available options make *A. baumannii* one of the most important topics of debate in clinical microbiology today. As demonstrated from our as well as others' data, colistin remains the single most effective antibiotic used in the treatment of MDR strains of *A. baumannii*. However, the emergence of colistin-resistant strains of the bacterium poses an extreme danger to healthcare systems worldwide. In a study published by Nowak et al., an alarmingly high incidence of pan-drug-resistant *A. baumannii* isolates were collected from ventilator-associated pneumonia in Greece, Italy, and Spain [43]. In their study, 45% of 65 clinically isolated strains were resistant to colistin. To overcome such complicated and severe infection, several enhancement methods, including increased colistin loading dose, higher maintenance dose, and combination therapy with other antibiotics, have been implemented [44]. As an example, the combination of colistin with nicodamids has been shown to be an effective treatment for *A. baumannii* colistin-resistant strains [45]. The interaction of nicodamids with the negatively charged outer membrane of colistin-resistant strains leads to a synergistic effect with colistin. Several other studies have suggested combination use of colistin with meropenem due to improvement in minimum inhibitory concentration of antibiotics in treatment of CR strains of the bacterium [46,47]. Despite the data demonstrating the benefit of drug combination therapy in cases of carbapenem- and colistin-resistant strains of the bacteria, recent studies tend to favour the single-treatment regimen or monotherapy due to its lower mortality rates and shorter length of stay (LOS), explained by the limited development of antibiotic-related adverse events. A meta-analysis published by Huang et al., demonstrated that colistin monotherapy was associated with similar rates of clinical improvement, drug-related toxicity, and hospital mortality in comparison to combination therapy with meropenem. They stated that combination therapy with meropenem was not superior to monotherapy with colistin. Similar data were recently published by other researchers. [47,48]. That is why the optimisation of treatment of such strains, especially due to their high mortality rate, and utilisation of best treatment method are widely open to debate and currently uncertain. As antibiotics are not part of chronic treatment regimens, large pharmaceutical companies have little interest in investing in new antibiotics [49]. Hence, drug repurposing strategies seem to be a better solution for the time being, even with their limitations. However, the combination of repurposed drugs with existing antibiotics may result in an increase of toxic side effects and interfere with the pharmacokinetics, pharmacodynamics, solubility, and conservation of the respective antibiotic [50]. That is why the need for new molecules is more urgent than ever. Until then, these results emphasise the extreme importance of routine environmental disinfection, staff hygiene, minimisation of invasive procedures or their implementation duration, optimisation of hospitalisation period in relation to age and comorbidities, and utilisation of a rational, strict antibiotic stewardship programme.

4. Materials and Methods

4.1. Healthcare Facilities, Patients, and Specimens

In this observational, prospective, cross-sectional study, we monitored and gathered data from all clinically relevant isolates of *A. baumannii* from patients hospitalised in two standard male and female wards, one an intermediary ward for critically ill patients with need of central monitoring and the other an intensive care unit of the fifth department of internal medicine of University Hospital Bratislava, Ružinov for a period of four years (from the beginning of 2019 to the end of 2022). The majority of the hospitalised patients were from the Bratislava region. To the best of our knowledge, our data (in whole or in part) were not used in production of any ECDC reports on antibiotic resistance of the bacterium during the study period. For each isolate, we collected information on place of sampling, antibiogram, whether the sample was collected from a HAI or non-healthcare associated infection (NHAI), and whether the patient was discharged or deceased during our monitoring. There was no specific process for the selection of patients in this study. However, all patients included in the study had active infectious processes caused by *A. baumannii* with raised inflammatory markers. Patients with a prior history of colonisation with *A. baumannii* were excluded from the study.

4.2. Isolation and Proof of Bacterial Strains

Sterile swabs were used to collect samples from the nose, tonsils, wounds, and decubitus. Blood sampling was performed via peripheral intravenous puncture. Urine was collected via mid-stream urine sampling or via urinary catheter. Samples were processed using conventional methods approved by the Slovak Ministry of Health. Strain identification was performed in accordance with classical isolation and classical bio-chemical and cultivation methods with a Bruker MALDI Biotyper (Bruker, Billerica, MA, USA), which uses the MALDI-TOF Mass Spectrometry method [51]. Antibiotic testing results were interpreted according to EUCAST guidelines (European Committee on Antimicrobial Susceptibility Testing). The panel of tested antibiotics was pre-determined by diagnostic laboratories, defined for Gram-negative isolates.

4.3. Statistical Analysis

Data collected during the monitoring period were analysed using Microsoft Excel. A simple linear regression test was used to calculate the significance of the antibiotic resistance trend over time. p value was determined using a t distribution with $n - 2$ degrees of freedom (df) (in a simple linear regression sample test). The confidence level of the p value was determined as 95%. P^1 as true prevalence was defined as the number of identified infected individuals with *A. baumannii* hospitalised in the department divided by the total number of hospitalised patients in the department for a given year. P^2 as CR prevalence was defined as the number of CR isolates of *A. baumannii* divided by the total number of hospitalised patients in the department for a given year. The combined resistant rate for *A. baumannii* was defined as the accumulated percentage of resistance of the bacterium to three categories of antibiotics (namely: fluoroquinolones, aminoglycosides, and carbapenems) in a given year. MDR was defined as the non-susceptibility of the bacterium to $\geq$1 agent in ≥3 antimicrobial categories. XDR was defined as non-susceptibility to ≥1 agent in all but ≤2 antimicrobial categories.

4.4. Ethical Consent

Ethical review and approval were not required for the study (as an observational study) on human participants in accordance with local legislation and institutional requirements. Written informed consent for participation was not required for this study in accordance with national legislation and the institutional requirements.

5. Conclusions

Multidrug resistance among all *A. baumannii* as well as extensive drug resistance among carbapenem-resistant strains of the bacterium has risen significantly over the last four years in our department. MDR is significantly associated with the prevalence of the bacterium, and with the current rising pattern of resistance, an increase in the prevalence of *A. baumannii* is expected in the coming years. Colistin, ampicillin-sulbactam, and to a much lesser extent gentamycin are the only antibiotics left to have efficacy on isolated strains of *A. baumannii*. No colistin-resistant strain of *A. baumannii* was isolated. Prolonged hospitalisation is associated with an increase in HAI rates. The significant association of HAI with the isolation of XDR strains, along with high rate of associated mortality of the bacterium, could be a predictor of poor outcome. These results emphasise the urgent need for new treatment options, optimisation of antibiotics stewardship programmes, and stricter epidemiologic control and management of the spread of the bacterium's infection.

Supplementary Materials: The following are available online at https://www.mdpi.com/article/10.3390/antibiotics12071200/s1, Supplementary File S1: Antibiotic resistant trends of *A. baumannii*. AMI: amikacin, AMP: ampicillin, ATM: aztreonam, CAZ: ceftazidime, CIP: ciprofloxacin, COL: colistin, COT: co-trimoxazole (trimethoprim-sulfamethoxazole), CTX: cefotaxime, CXM: cefuroxime, ETP: ertapenem, FEP: cefepime, GEN: gentamycin, IMI: imipenem, MEM: meropenem, SAM: ampicillin-sulbactam, SPZ: sulperazone (cefoperazone-sulbactam), TET: tetracycline, TOB: tobramycin, TZP: tazocin (piperacillin-tazobactam).

Author Contributions: Conceptualisation, Y.J. and M.J.; data curation, Y.J. and M.J.; formal analysis, Y.J.; methodology, Y.J. and A.L.; investigation, Y.J. and M.J.; software, M.J. and Y.J.; supervision, A.L. and J.P.; visualisation, Y.J.; writing—original draft, Y.J. and M.J.; writing—review and editing, Y.J., A.L. and J.P. All authors have read and agreed to the published version of the manuscript.

Funding: This research received no external funding.

Institutional Review Board Statement: Not applicable.

Informed Consent Statement: Not applicable.

Data Availability Statement: Not applicable.

Conflicts of Interest: The authors declare no conflict of interest.

eferences

Ma, C.; McClean, S. Mapping Global Prevalence of Acinetobacter baumannii and Recent Vaccine Development to Tackle It. *Vaccines* **2021**, *9*, 570. [CrossRef]

Almasaudi, S.B. *Acinetobacter* spp. as nosocomial pathogens: Epidemiology and resistance features. *Saudi J. Biol. Sci.* **2018**, *25*, 586–596. [CrossRef] [PubMed]

Fournier, P.E.; Richet, H.; Weinstein, R.A. The Epidemiology and Control of Acinetobacter baumannii in Health Care Facilities. *Clin. Infect. Dis.* **2006**, *42*, 692–699. [CrossRef] [PubMed]

Yashar, J.; Monika, J.; Juraj, P. Acinetobacter baumannii: Emergence of a Superbug, Past, Present, and Future. In *The Global Antimicrobial Resistance Epidemic*; Guillermo, T.-I., Ed.; IntechOpen: Rijeka, Croatia, 2022; Chapter 13.

Ko, S.-Y.; Kim, N.; Park, S.-Y.; Kim, S.-Y.; Shin, M.; Lee, J.-C. Acinetobacter baumannii under Acidic Conditions Induces Colistin Resistance through PmrAB Activation and Lipid A Modification. *Antibiotics* **2023**, *12*, 813. [CrossRef] [PubMed]

Nasr, P. Genetics, epidemiology, and clinical manifestations of multidrug-resistant Acinetobacter baumannii. *J. Hosp. Infect.* **2020**, *104*, 4–11. [CrossRef]

7. Vázquez-López, R.; Solano-Gálvez, S.G.; Juárez Vignon-Whaley, J.J.; Abello Vaamonde, J.A.; Padró Alonzo, L.A.; Rivera Resénd A.; Muleiro Álvarez, M.; Vega López, E.N.; Franyuti-Kelly, G.; Álvarez-Hernández, D.A.; et al. Acinetobacter baumann Resistance: A Real Challenge for Clinicians. *Antibiotics* **2020**, *9*, 205. [CrossRef]
8. European Centre for Disease Prevention and Control. *Surveillance Atlas of Infectious Diseases*; ECDC: Solna, Sweden, 2021.
9. Novović, K.; Jovčić, B. Colistin Resistance in Acinetobacter baumannii: Molecular Mechanisms and Epidemiology. *Antibioti* **2023**, *12*, 516. [CrossRef]
10. Arefian, H.; Hagel, S.; Fischer, D.; Scherag, A.; Brunkhorst, F.M.; Maschmann, J.; Hartmann, M. Estimating extra length of st due to healthcare-associated infections before and after implementation of a hospital-wide infection control program. *PLoS ON* **2019**, *14*, e0217159. [CrossRef] [PubMed]
11. Jiang, J.; Upfill-Brown, A.; Dann, A.M.; Kim, S.S.; Girgis, M.D.; King, J.C.; Donahue, T.R. Association of Hospital Length of St and Complications With Readmission After Open Pancreaticoduodenectomy. *JAMA Surg.* **2019**, *154*, 88–90. [CrossRef]
12. Sulzgruber, P.; Schnaubelt, S.; Koller, L.; Laufer, G.; Pilz, A.; Kazem, N.; Winter, M.-P.; Steinlechner, B.; Andreas, M.; Fleck, et al. An Extended Duration of the Pre-Operative Hospitalization is Associated with an Increased Risk of Healthcare-Associate Infections after Cardiac Surgery. *Sci. Rep.* **2020**, *10*, 8006. [CrossRef] [PubMed]
13. Jalali, Y.; Sturdik, I.; Jalali, M.; Payer, J. Isolated carbapenem resistant bacteria, their multidrug resistant profile, percentage healthcare associated infection and associated mortality, in hospitalized patients in a University Hospital in Bratislava. *Brati Lek. Listy* **2021**, *122*, 379–385. [CrossRef]
14. Beavers, S.F.; Blossom, D.B.; Wiemken, T.L.; Kawaoka, K.Y.; Wong, A.; Goss, L.; McCormick, M.I.; Thoroughman, D.; Srinivasa A. Comparison of risk factors for recovery of Acinetobacter baumannii during outbreaks at two Kentucky hospitals, 2006 *Pub Health Rep.* **2009**, *124*, 868–874. [CrossRef]
15. Sheng, W.-H.; Liao, C.-H.; Lauderdale, T.-L.; Ko, W.-C.; Chen, Y.-S.; Liu, J.-W.; Lau, Y.-J.; Wang, L.-S.; Liu, K.-S.; Tsai, T.-Y.; et al. multicenter study of risk factors and outcome of hospitalized patients with infections due to carbapenem-resistant Acinetobact baumannii. *Int. J. Infect. Dis.* **2010**, *14*, e764–e769. [CrossRef] [PubMed]
16. Alrahmany, D.; Omar, A.F.; Alreesi, A.; Harb, G.; Ghazi, I.M. *Acinetobacter baumannii* Infection-Related Mortality in Hospitalize Patients: Risk Factors and Potential Targets for Clinical and Antimicrobial Stewardship Interventions. *Antibiotics* **2022**, *11*, 108 [CrossRef]
17. Ellis, D.; Cohen, B.; Liu, J.; Larson, E. Risk factors for hospital-acquired antimicrobial-resistant infection caused by Acinetobact baumannii. *Antimicrob. Resist. Infect. Control* **2015**, *4*, 40. [CrossRef] [PubMed]
18. Chusri, S.; Chongsuvivatwong, V.; Silpapojakul, K.; Singkhamanan, K.; Hortiwakul, T.; Charernmak, B.; Doi, Y. Clinic characteristics and outcomes of community and hospital-acquired Acinetobacter baumannii bacteremia. *J. Microbiol. Immun Infect.* **2019**, *52*, 796–806. [CrossRef] [PubMed]
19. Mehta, Y.; Gupta, A.; Todi, S.; Myatra, S.; Samaddar, D.P.; Patil, V.; Bhattacharya, P.K.; Ramasubban, S. Guidelines for preventic of hospital acquired infections. *Indian J. Crit. Care Med.* **2014**, *18*, 149–163. [CrossRef]
20. Zanetti, G.; Blanc, D.S.; Federli, I.; Raffoul, W.; Petignat, C.; Maravic, P.; Francioli, P.; Berger, M.M. Importation of Acinetobact baumannii Into a Burn Unit: A Recurrent Outbreak of Infection Associated With Widespread Environmental Contaminatic *Infect. Control Hosp. Epidemiol.* **2007**, *28*, 723–725. [CrossRef]
21. Blanco, N.; Harris, A.D.; Rock, C.; Johnson, J.K.; Pineles, L.; Bonomo, R.A.; Srinivasan, A.; Pettigrew, M.M.; Thom, K.A. Ris Factors and Outcomes Associated with Multidrug-Resistant Acinetobacter baumannii upon Intensive Care Unit Admissic *Antimicrob. Agents Chemother.* **2018**, *62*, e01631-17. [CrossRef]
22. Zeighami, H.; Valadkhani, F.; Shapouri, R.; Samadi, E.; Haghi, F. Virulence characteristics of multidrug resistant biofilm formir Acinetobacter baumannii isolated from intensive care unit patients. *BMC Infect. Dis.* **2019**, *19*, 629. [CrossRef]
23. Ababneh, Q.; Abulaila, S.; Jaradat, Z. Isolation of extensively drug resistant Acinetobacter baumannii from environmental surfac inside intensive care units. *Am. J. Infect. Control* **2022**, *50*, 159–165. [CrossRef] [PubMed]
24. Ceparano, M.; Baccolini, V.; Migliara, G.; Isonne, C.; Renzi, E.; Tufi, D.; De Vito, C.; De Giusti, M.; Trancassini, M.; Alessandri, et al. Acinetobacter baumannii Isolates from COVID-19 Patients in a Hospital Intensive Care Unit: Molecular Typing and Ris Factors. *Microorganisms* **2022**, *10*, 722. [CrossRef]
25. Sirijatuphat, R.; Thawornkaew, S.; Ruangkriengsin, D.; Thamlikitkul, V. Colistin Monotherapy versus Colistin plus Sitafloxacin f Therapy of Carbapenem-Resistant Acinetobacter baumannii Infections: A Preliminary Study. *Antibiotics* **2022**, *11*, 1707. [CrossRe [PubMed]
26. Kogilathota Jagirdhar, G.S.; Rama, K.; Reddy, S.T.; Pattnaik, H.; Qasba, R.K.; Elmati, P.R.; Kashyap, R.; Schito, M.; Gupta, Efficacy of Cefoperazone Sulbactam in Patients with Acinetobacter Infections: A Systematic Review of the Literature. *Antibioti* **2023**, *12*, 582. [CrossRef] [PubMed]
27. Ungthammakhun, C.; Vasikasin, V.; Changpradub, D. A Randomized Controlled Trial of Colistin Combined with Sulbactam: 9 per Day versus 12 g per Day in the Treatment of Extensively Drug-Resistant Acinetobacter baumannii Pneumonia: An Interi Analysis. *Antibiotics* **2022**, *11*, 1112. [CrossRef]
28. European Centre for Disease Prevention and Control. *Antimicrobial Consumption in the EU/EEA (ESAC-Net)—Annual Epidemiologic Report for 2021*; ECDC: Solna, Sweden, 2021; p. 28.

29. European Centre for Disease Prevention and Control. Antimicrobial resistance in the EU/EEA (EARS-Net). In *Yearly*, 2022th ed.; European Centre for Disease Prevention and Control: Solna, Sweden, 2022; p. 36.
30. Tacconelli, E.; Carrara, E.; Savoldi, A.; Harbarth, S.; Mendelson, M.; Monnet, D.L.; Pulcini, C.; Kahlmeter, G.; Kluytmans, J.; Carmeli, Y.; et al. Discovery, research, and development of new antibiotics: The WHO priority list of antibiotic-resistant bacteria and tuberculosis. *Lancet Infect. Dis.* **2018**, *18*, 318–327. [CrossRef]
31. Mabrouk, S.S.; Abdellatif, G.R.; El-Ansary, M.R.; Aboshanab, K.M.; Ragab, Y.M. Carbapenemase Producers Among Extensive Drug-Resistant Gram-Negative Pathogens Recovered from Febrile Neutrophilic Patients in Egypt. *Infect. Drug Resist.* **2020**, *13*, 3113–3124. [CrossRef]
32. Naik, P.; Pandey, S.; Gagan, S.; Biswas, S.; Joseph, J. Virulence factors in multidrug (MDR) and Pan-drug resistant (XDR) Pseudomonas aeruginosa: A cross-sectional study of isolates recovered from ocular infections in a high-incidence setting in southern India. *J. Ophthalmic Inflamm. Infect.* **2021**, *11*, 36. [CrossRef]
33. Appaneal, H.J.; Lopes, V.V.; LaPlante, K.L.; Caffrey, A.R. Treatment, Clinical Outcomes, and Predictors of Mortality among a National Cohort of Admitted Patients with *Acinetobacter baumannii* Infection. *Antimicrob. Agents Chemother.* **2022**, *66*, e0197521. [CrossRef]
34. Wachino, J.I.; Jin, W.; Kimura, K.; Arakawa, Y. Intercellular Transfer of Chromosomal Antimicrobial Resistance Genes between Acinetobacter baumannii Strains Mediated by Prophages. *Antimicrob. Agents Chemother.* **2019**, *63*, e00334-19. [CrossRef]
35. Leungtongkam, U.; Thummeepak, R.; Tasanapak, K.; Sitthisak, S. Acquisition and transfer of antibiotic resistance genes in association with conjugative plasmid or class 1 integrons of Acinetobacter baumannii. *PLoS ONE* **2018**, *13*, e0208468. [CrossRef]
36. Semenec, L.; Cain, A.K.; Dawson, C.J.; Liu, Q.; Dinh, H.; Lott, H.; Penesyan, A.; Maharjan, R.; Short, F.L.; Hassan, K.A.; et al. Cross-protection and cross-feeding between Klebsiella pneumoniae and Acinetobacter baumannii promotes their co-existence. *Nat. Commun.* **2023**, *14*, 702. [CrossRef] [PubMed]
37. Vijayakumar, S.; John, J.J.; Vasudevan, K.; Mathur, P.; Ray, P.; Neeravi, A.; Baskaran, A.; Kirubananthan, A.; Anandan, S.; Biswas, I.; et al. Genomic rearrangements of mobile genetic elements associated with carbapenem resistance of *Acinetobacter baumannii*. *bioRxiv* **2022**. [CrossRef]
38. Wyres, K.L.; Holt, K.E. Klebsiella pneumoniae as a key trafficker of drug resistance genes from environmental to clinically important bacteria. *Curr. Opin. Microbiol.* **2018**, *45*, 131–139. [CrossRef]
39. Lam, M.M.C.; Wick, R.R.; Wyres, K.L.; Gorrie, C.L.; Judd, L.M.; Jenney, A.W.J.; Brisse, S.; Holt, K.E. Genetic diversity, mobilisation and spread of the yersiniabactin-encoding mobile element ICEKp in Klebsiella pneumoniae populations. *Microb. Genom.* **2018**, *4*, e000196. [CrossRef]
40. Koreň, J.; Andrezál, M.; Drahovská, H.; Hubenáková, Z.; Liptáková, A.; Maliar, T. Next-Generation Sequencing of Carbapenem-Resistant Klebsiella pneumoniae Strains Isolated from Patients Hospitalized in the University Hospital Facilities. *Antibiotics* **2022**, *11*, 1538. [CrossRef] [PubMed]
41. Koren, J.; Hubenakova, Z.; Drahovska, H.; Ozaee, E.; Markuskova, B.; Lichvarikova, A. Emergence of extended-spectrum β-lactamase (ESBL) and/or carbapenemase producing Enterobacteriaceae (CPE) and their antimicrobial resistance. *Bratisl. Lek. Listy* **2019**, *120*, 935–940. [CrossRef]
42. Nowak, J.; Zander, E.; Stefanik, D.; Higgins, P.G.; Roca, I.; Vila, J.; McConnell, M.J.; Cisneros, J.M.; Seifert, H.; WP4, M.W.G. High incidence of pandrug-resistant Acinetobacter baumannii isolates collected from patients with ventilator-associated pneumonia in Greece, Italy and Spain as part of the MagicBullet clinical trial. *J. Antimicrob. Chemother.* **2017**, *72*, 3277–3282. [CrossRef] [PubMed]
43. Huang, C.; Chen, I.; Tang, T. Colistin Monotherapy versus Colistin plus Meropenem Combination Therapy for the Treatment of Multidrug-Resistant Acinetobacter baumannii Infection: A Meta-Analysis. *J. Clin. Med.* **2022**, *11*, 3239. [CrossRef]
44. Ayerbe-Algaba, R.; Gil-Marques, M.L.; Jimenez-Mejias, M.E.; Sanchez-Encinales, V.; Parra-Millan, R.; Pachon-Ibanez, M.E.; Pachon, J.; Smani, Y. Synergistic Activity of Niclosamide in Combination With Colistin Against Colistin-Susceptible and Colistin-Resistant Acinetobacter baumannii and Klebsiella pneumoniae. *Front. Cell. Infect. Microbiol.* **2018**, *8*, 348. [CrossRef] [PubMed]
45. Gunalan, A.; Sarumathi, D.; Sastry, A.S.; Ramanathan, V.; Rajaa, S.; Sistla, S. Effect of combined colistin and meropenem against meropenem resistant Acinetobacter baumannii and Pseudomonas aeruginosa by checkerboard method: A cross sectional analytical study. *Indian J. Pharmacol.* **2021**, *53*, 207–212. [CrossRef] [PubMed]
46. Katip, W.; Uitrakul, S.; Oberdorfer, P. A Comparison of Colistin versus Colistin Plus Meropenem for the Treatment of Carbapenem-Resistant Acinetobacter baumannii in Critically Ill Patients: A Propensity Score-Matched Analysis. *Antibiotics* **2020**, *9*, 647. [CrossRef] [PubMed]
47. Kaye, K.S.; Marchaim, D.; Thamlikitkul, V.; Carmeli, Y.; Chiu, C.-H.; Daikos, G.; Dhar, S.; Durante-Mangoni, E.; Gikas, A.; Kotanidou, A.; et al. Colistin Monotherapy versus Combination Therapy for Carbapenem-Resistant Organisms. *NEJM Evid.* **2023**, *2*, EVIDoa2200131. [CrossRef]
48. Peyclit, L.; Baron, S.A.; Rolain, J.M. Drug Repurposing to Fight Colistin and Carbapenem-Resistant Bacteria. *Front. Cell. Infect. Microbiol.* **2019**, *9*, 193. [CrossRef]
49. Sun, W.; Sanderson, P.E.; Zheng, W. Drug combination therapy increases successful drug repositioning. *Drug Discov. Today* **2016**, *21*, 1189–1195. [CrossRef] [PubMed]

50. Merck. MALDI-MS Analysis of Proteins. 2020. Available online: https://www.sigmaaldrich.com/life-science/proteomics/recombinant-protein-expression/protocols/maldims-analysis-of-proteins.html?gclid=Cj0KCQiAk53-BRD0ARIsAJuNhpvUwYCIMd5NxDbG8_w2Btapg0Mg0Ixf1ueMvjLxZgSP302NkYcR6CQaApLTEALw_wcB (accessed on 20 July 2020).
51. EUCAST. *Clinical Breakpoints—Breakpoints and Guidance*; EUCAST: Copenhagen, Denmark, 2023.

antibiotics

MDPI

rticle

creening for Metallo-Beta-Lactamases Using Non-Carbapenem gents: Effective Detection of MBL-Producing *Enterobacterales* nd Differentiation of Carbapenem-Resistant *Enterobacterales*

entarou Takei [1,*], Hajime Kanamori [1], Asami Nakayama [2], Mikiko Chiba [2], Yumiko Takei [1], Issei Seike [1], hiho Kitamura [1], Hiroaki Baba [1], Kengo Oshima [1] and Koichi Tokuda [1]

1 Department of Infectious Diseases, Internal Medicine, Tohoku University Graduate School of Medicine, Sendai 980-8574, Japan; kanamori@med.tohoku.ac.jp (H.K.); chiho.kitamura.d1@tohoku.ac.jp (C.K.); koshima@med.tohoku.ac.jp (K.O.)

2 Department of Laboratory Medicine, Tohoku University Hospital, Sendai 980-8574, Japan

* Correspondence: kentarou.takei.e7@med.tohoku.ac.jp; Tel.: +81-22-717-7373

Abstract: Metallo-beta-lactamases (MBLs) are enzymes that break down carbapenem antibiotics, leading to carbapenem-resistant organisms. Carbapenemase-resistant *Enterobacterales* (CRE) is one of them. Outbreaks of CRE infection can occur in healthcare facilities and lead to increased deaths, illness, and medical costs. This study was conducted to detect MBLs using non-carbapenem agents and exclude MBLs among CRE isolates. A total of 3776 non-duplicate sequential *Enterobacterales* isolates from a single facility were screened between January 2019 and December 2022 using non-carbapenem agents, ceftazidime and cefoperazone/sulbactam. Positive 153 isolates (4.0%) were further tested using carbapenemase-confirmation tests and verified through polymerase chain reaction (PCR) testing. Fifteen imipenemase (IMP)-type MBL-producing *Enterobacterales* (0.4%) including one susceptible to carbapenems were identified. Moreover, 160 isolates (4.2%) meeting the criteria for CRE were directly subjected to PCR testing. All fourteen CRE isolates with MBLs identified through PCR testing were found to be the same strains screened using ceftazidime and cefoperazone/sulbactam. Screening using ceftazidime and cefoperazone/sulbactam can effectively detect MBL-producing *Enterobacterales* strains. This screening method showed comparable results to screening with meropenem, potentially serving as a supplementary approach and contributing to differentiating between MBL- and non-MBL-producing CRE strains. Our findings support these screening methods, particularly in regions where IMP-type MBLs are prevalent.

Keywords: ceftazidime; cefoperazone/sulbactam; carbapenem-resistant *Enterobacterales*; metallo-β-lactamase

1. Introduction

Metallo-beta-lactamase (MBL) is a type of carbapenemase enzyme that breaks down carbapenem antibiotics. MBLs are zinc-type β-lactamases categorized under Class B according to the Ambler classification, the most widely used classification of β-lactamases [1]. *Enterobacterales* that produce carbapenemase are referred to as carbapenemase-producing *Enterobacterales* (CPE). Clinical MBL-producing *Enterobacterales* was first reported in 1994 in a *Serratia marcescens* strain resistant to imipenem [2]. Resistance of MBL-producing *Enterobacterales* to carbapenem drugs is of particular concern as they have the potential to horizontally transmit resistance to other *Enterobacterales* via conjugative transfer mediated by plasmids [3]. Following the emergence of new types of carbapenemase-producing bacteria in the United States, such as the *Klebsiella pneumoniae* carbapenemase-producing strains outbreak in 1996 and the New Delhi metallo-beta-lactamase-producing strains in 2009, the Centers for Disease Control and Prevention issued a warning about carbapenem-resistant *Enterobacterales* (CRE) in 2013, drawing attention to the presence of CRE [4–6].

Outbreaks of CRE infection frequently occur in healthcare facilities and lead to increased deaths, illness, and medical costs [7]. In Japan, the definition of CRE within the country was established in September 2014 under the Infectious Diseases Control Law, which includes the presence of *Enterobacterales* isolates with a minimum inhibitory concentration (MIC) $\geq$2 μg/mL for meropenem, $\geq$2 μg/mL for imipenem, and $\geq$64 μg/mL for cefmetazole. As CRE is defined by resistance to carbapenem drugs, it encompasses various mechanisms of resistance, including resistance due to MBLs. CPEs account for 16.5–28% of CRE isolates; in Japan, most of these CPEs were reported to have imipenemases (IMP)-type MBLs [8]. However, MBL-producing *Enterobacterales* do not necessarily correspond to CRE, as those strains may be susceptible to carbapenem.

Representative clinical laboratory organizations, such as the Clinical and Laboratory Standards Institute (CLSI) or European Committee on Antimicrobial Susceptibility Testing (EUCAST), have proposed as methods for CPE screening [9,10]. A common aspect of these testing methods is the evaluation of drug susceptibility, particularly toward carbapenem agents such as meropenem. Nishio et al. [11] conducted a survey of MBL-producing Gram-negative rods from 13 clinical laboratories in a specific region of Japan, in 2004, ten years before the Infectious Diseases Control Law regarding CRE was established in Japan. As a first-step screening for the collected bacterial strains, drug susceptibility testing using ceftazidime and cefoperazone/sulbactam was employed. Then, confirmation of carbapenemase production in drug-resistant strains was carried out using carbapenem confirmation tests and PCR examinations, as reported in the study. In our facility, carbapenemase confirmation tests were previously conducted on suspicious MBL-producing *Enterobacterales*, following preset protocols provided by the manufacturer. These protocols involved assessing non-susceptibility to third- or fourth-generation cephalosporins and carbapenem agents. Additionally, empirical judgment was employed by observing resistance to imipenem. However, this method often imposed a large burden in terms of the laboratory workload. After that, focusing on screening with ceftazidime and cefoperazone/sulbactam, we have been implementing MBL screening methods with these two drugs for some time as part of the testing protocol.

In this study, we evaluated the usefulness of screening for MBLs using non-carbapenem antimicrobial agents, ceftazidime and cefoperazone/sulbactam, against *Enterobacterales*. Drug susceptibility testing for ceftazidime and cefoperazone/sulbactam among CRE, facilitated by automated antimicrobial susceptibility testing, demonstrates the ability to rapidly identify non-MBL-producing CRE.

2. Results

2.1. Total Bacterial Isolates and Those Screened Using Ceftazidime and Cefoperazone/Sulbactam

A total of 3776 non-duplicate sequential strains were obtained (Table S1). Of these, a total of 153 isolates (4.0%) were screened using ceftazidime and cefoperazone/sulbactam. A list of those strains is shown in Table 1.

2.2. Enterobacterales Screened Using Ceftazidime and Cefoperazone/Sulbactam

Of the 153 isolates screened using ceftazidime and cefoperazone/sulbactam, *Enterobacter cloacae* was predominant, accounting for the highest proportion (40.5%) within this population, followed by *K. pneumoniae* (15.7%), *Escherichia coli* (15.0%), and *Citrobacter braakii* (11.8%). *Klebsiella aerogenes* was observed in only four cases. Of the 153 isolates, 15 (9.8%) were selected after the sodium mercaptoacetic acid (SMA) test (Figure 1). These fifteen strains tested positive in PCR examination, and all were identified as IMP-1 type MBL. Of these, fourteen MBL-possessing strains resistant to meropenem were observed, which corresponded to CRE. However, one isolate was identified as carbapenem-sensitive MBL, showing susceptibility to both meropenem and imipenem. The MBLs-producing isolates are listed in Table 2.

Table 1. List of *Enterobacterales* resistant to both ceftazidime and cefoperazone/sulbactam among the first cultured non-duplicate sequential isolates.

***Enterobacterales* Resistant to Ceftazidime and Cefoperazone/Sulbactam**	**n = 153**
Enterobacter cloacae	62
Klebsiella pneumoniae	24
Escherichia coli	23
Citrobacter braakii	18
Serratia marcescens	6
Klebsiella aerogenes	4
Citrobacter sp.	3
Klebsiella oxytoca	2
Morganella morganii	2
Proteus mirabilis	2
Serratia liquefaciens	2
Citrobacter werkmanii	1
Citrobacter youngae	1
Enterobacter sp.	1
Hafnia alvei	1
Yersinia enterocolitica	1

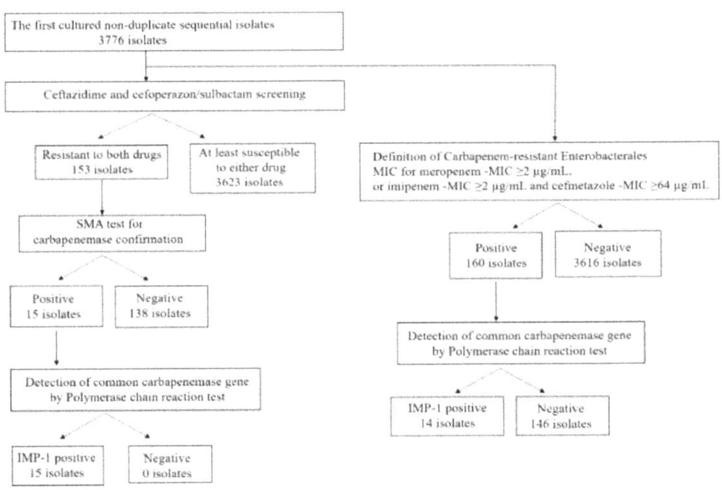

Figure 1. Identification and exclusion process for isolates screened using ceftazidime and cefoperazone/sulbactam, based on the definition of carbapenem-resistant *Enterobacterales*. Abbreviations: IMP, imipenemase; SMA, sodium mercaptoacetic acid; MIC, minimum inhibitory concentration.

2.3. Carbapenem-Resistant Enterobacterales

A total of 160 isolates were identified as CREs, corresponding to approximately 4.2% of the initial 3776 strains (Table 3). Among these isolates, 5 (3.1%) were resistant only to meropenem, 140 (87.5%) were resistant only to imipenem, and 15 (9.4%) were resistant to meropenem and imipenem. The major strains of CRE were *K. aerogenes* and *E. cloacae* (Table 3). The proportions of identified CRE strains among *K. aerogenes* and *E. cloacae* were 44.2% (96/217) and 12.4% (54/435), respectively. All CRE of *K. aerogenes* lacked MBLs and were resistant only to imipenem. Excluding isolates possessing MBLs, 95% (42/44) of *E. cloacae* showed resistance only to imipenem. Among the CRE strains, 14 strains (8.8%) produced MBLs, as confirmed via PCR testing. Among the CRE isolates that exhibited resistance solely to imipenem, none had MBLs; however, all MBL-producing CRE strains were resistant to meropenem.

Table 2. Characteristics and drug susceptibility of MBL-producing isolates.

No.	Bacterial Strains	CRE or Non-CRE	Sample	Antibiotic Minimum Inhibitory Concentration (μg/mL) MEM	IPM	CMZ	CAZ	CFP/SBT	CRO	FEP	LVX	TZP	GM
1	*Enterobacter cloacae*	CRE	Urine	>2	2	>32	>8	>32	>2	>16	1	≤4	≤2
2	*Enterobacter cloacae*	CRE	Sputum	>2	>2	>32	>8	>32	>2	>16	4	64	≤2
3	*Enterobacter cloacae*	CRE	Blood	2	≤0.5	>32	>8	32	>2	>16	1	8	≤2
4	*Enterobacter cloacae*	CRE	Blood	>2	>2	>32	>8	>32	>2	16	1	4	≤2
5	*Enterobacter cloacae*	CRE	Urine	>2	>2	>32	>8	>32	>2	16	≤0.12	16	≤2
6	*Enterobacter cloacae*	CRE	Urine	>2	>2	>32	>8	>32	>2	16	1	>64	≤2
7	*Enterobacter cloacae*	CRE	Urine	2	1	>32	>8	>32	>2	8	≤0.12	≤4	≤2
8	*Enterobacter cloacae*	CRE	Sputum	>2	1	>32	>8	32	>2	8	1	16	≤2
9	*Enterobacter cloacae*	CRE	Urine	>2	2	>32	>8	>32	>2	>16	2	>64	≤2
11	*Enterobacter cloacae*	CRE	Sputum	>2	2	>32	>8	>32	>2	>16	1	≤4	≤2
10	*Enterobacter cloacae*	Non-CRE	Sputum	≤0.25	≤0.5	>32	>8	>32	>2	8	1	16	≤2
12	*Klebsiella pneumoniae*	CRE	Bile	>2	>2	>32	>8	>32	>2	>16	4	64	≤2
13	*Klebsiella pneumoniae*	CRE	Puncture fluid	>2	>2	>32	>8	>32	>2	>16	≤0.12	64	≤2
14	*Klebsiella pneumoniae*	CRE	Urine	>2	>2	>32	>8	>32	>2	>16	1	>64	≤2
15	*Escherichia coli*	CRE	Urine	2	2	>32	>8	32	>2	4	1	≤4	≤2

Abbreviations: MEM, meropenem; IPM, imipenem; CMZ, cefmetazole; CAZ, ceftazidime; SBT/CFP, cefoperazone and sulbactam; CRO, ceftriaxone; FEP, cefepime; LVX, levofloxacin; TZP, tazobactam/piperacillin; GM, gentamicin; MBL, metallo-beta-lactamase; CRE, carbapenemase-resistant *Enterobacterales*.

Table 3. Carbapenem-resistant *Enterobacterales* among 3776 strains between January 2019 and December 2022.

Carbapenem-Resistant *Enterobacterales*	n = 160
Klebsiella aerogenes	96
Enterobacter cloacae	54
Klebsiella pneumoniae	8
Escherichia coli	2

2.4. Carbapenem, Ceftazidime, and Cefoperazone/Sulbactam-Resistant Enterobacterales

All MBL-producing CREs resistant to both ceftazidime and cefoperazone/sulbactam (Table 2). Therefore, in cases where susceptibility was demonstrated to either of the two drugs, potential MBL-producing CREs could be ruled out with 100% sensitivity and with statistical significance ($p < 0.001$) using Fisher's exact test (Table S2).

3. Discussion

We focused on two main aspects: screening of MBLs-producing *Enterobacterales* using non-carbapenem agents and exploring the potential of using drug susceptibility to ceftazidime and cefoperazone/sulbactam in order to rapidly exclude MBL-producing *Enterobacterales* among CRE isolates.

We screened on 3776 clinical specimens using ceftazidime and cefoperazone/sulbactam and identified 1 isolates of carbapenem-susceptible MBL. The EUCAST method is a highly effective approach and is likely to become a standard indicator; however, even with CPE screening based on the EUCAST criteria, 1.6% cases cannot be screened [12]. This study is a rare investigation focusing on the screening of MBLs using non-carbapenem agents. Screening with ceftazidime and cefoperazone/sulbactam proved to be a method as equally effective as screening with meropenem, at least with an MIC for meropenem >0.25 μg/mL; however, it is unknown whether our isolate of carbapenem-susceptible MBL indicating the MIC for meropenem ≤0.25 μg/mL can be screened using the EUCAST criteria due to lack of MIC details for meropenem.

CLSI and EUCAST, as world-renowned testing organizations, have proposed breakpoints for screening CPE. Each organization recommends conducting carbapenemase confirmation tests when the MIC of imipenem is ≥2 μg/mL and that of meropenem is ≥0.25 μg/mL when screening for CPE [9,10,13]. However, susceptibility to meropenem is given greater emphasis over that to imipenem [9,10]. Indeed, in our study, no MBL-producing CREs were identified through screening using only imipenem. Hence, screening for CPE using imipenem may have low utility potential. It was reported that at the meropenem screening breakpoint provided by CLSI (≥2 μg/mL), 86.2% of CPE isolates would have been detected. In contrast, the EUCAST screening breakpoint for meropenem (≥0.25 μg/mL) resulted in detection of 98.4% of CPE isolates [12]. The meropenem screening breakpoint based on EUCAST criteria, which is also included in the joint proposal from four domestic academic societies (Japanese Society of Chemotherapy, Japanese Association for Infectious Diseases, Japanese Society for Infection Prevention and Control, and Japanese Society for Clinical Microbiology), was suggested for screening MBLs [14]. However, while these criteria are academically recommended, it is unclear whether panels satisfying these recommendations are widely being used in actual clinical laboratories.

This study also investigated MBLs from the perspective of CRE. Among the CRE strains, *K. aerogenes* accounted for 59.6%, followed by *E. cloacae*, which is consistent with previous reports in Japan [8]. In our study, all *K. aerogenes* strains were resistant solely to imipenem and non-MBL-producers. Even in *E. cloacae*, 95% of non-MBL-producing CREs were resistant only to imipenem. Among the non-MBL-producing CRE strains, CRE strains of *K. aerogenes* and *E. cloacae* have developed via mechanisms such as AmpC overexpression and efflux pumps, and downregulation of porin is involved in the resistance mechanisms, including the coexistence of extended-spectrum β-lactamase (ESBLs) [15–18]. Screening based on the definition of CRE more frequently detects AmpC-producing bacteria (e.g., *K. aerogenes* and *E. cloacae*). However, screening with ceftazidime and cefoperazone/sulbactam tends to be more effective in detecting ESBL-producing bacteria (e.g., *E. coli* and *K. pneumoniae*) [19,20]. Finally, screening with ceftazidime and cefoperazone/sulbactam excluded 98.2% (213/217) of *K. aerogenes* and 85.7% (373/435) of *E. cloacae*, which are representative AmpC-producing bacteria. Additionally, it excluded 98.2% (1240/1263) of *E. coli* and 96.3% (629/653) of *K. pneumoniae*, which are representative ESBLs, through screening.

In our current study, 14 isolates of MBL-producing CREs showed resistance to ceftazidime and cefoperazone/sulbactam. In contrast, no MBLs were detected using PCR testing in the CRE strains that exhibited susceptibility to either ceftazidime or cefoperazone/sulbactam. Based on these findings, all CRE strains demonstrating susceptibility to ceftazidime or cefoperazone/sulbactam may be non-MBL strains, as confirmed through statistical analyses. This finding suggests the possibility of MBLs being excluded, potentially contributing to early infection control based on automated drug susceptibility testing. It is important to differentiate CPE from non-CPE strains among CRE because the CPE acquisition is associated with horizontal transmission, whereas non-carbapenemase-producing CRE is associated with carbapenem exposure. Differences in the acquisition factors necessitate tailored infection prevention efforts [21].

This study has several limitations. The molecular epidemiology of carbapenemase varies widely worldwide, and the prevalence of IMP-type MBL is centered around east Asia or other countries, including Japan [22]. Particularly, the IMP type is predominant in Japan [8]. The molecular epidemiology differs from that in other countries [23]. We focused on bacterial strains from a single facility in a specific region; hence, our results may not reflect the global situation. In addition, it is unclear whether screening with ceftazidime and cefoperazone/sulbactam would be effective against other CPE strains because all of our MBLs were of the IMP type. Second, carbapenem confirmation tests were not performed on all the strains that were excluded during the screening process of ceftazidime and cefoperazone/sulbactam. Therefore, we could not rule out the possibility of the presence of MBLs or other CPE strains that are susceptible to carbapenem among the negative strains identified through screening with these two agents.

4. Materials and Methods

4.1. Study Design for Bacterial Isolates

A total of 3776 non-duplicate sequential *Enterobacterales* isolates were cultured and identified from various clinical specimens at Tohoku University Hospital (1160 beds) located in Sendai, which is in the northeastern part of Japan, between January 2019 and December 2022. These *Enterobacterales* strains were identified using a VITEK® MS (Sysmex-bioMérieux, Tokyo, Japan). Only the first 3839 cultured non-duplicate sequential isolates were retrospectively selected. These isolates were initially screened based on their drug susceptibility to ceftazidime and cefoperazone/sulbactam. Isolates that screened positive underwent further screening using the SMA method, which is a well-known carbapenemase confirmation test [24]. Finally, PCR testing was performed on strains that tested positive in the SMA method. However, within the first cultured non-duplicate sequential isolates, bacteria that met the criteria for CRE were selected and directly tested using PCR. The results of PCR testing were compared with those of the screening method using ceftazidime and cefoperazone/sulbactam.

4.2. Antimicrobial Susceptibility Testing

Drug susceptibility testing of the bacterial isolates was performed using Microscan WalkAway® (Beckman Coulter, Brea, CA, USA) and accompanying panels of the Microscan Neg® series (Beckman Coulter, Brea, CA, USA). The bacterial solution used for this test was mainly prepared using the prompt inoculation method, but the standard inoculation method (CLSI method) was used for preparing potential CRE isolates or blood specimens [25–27]. Because we used a previous commercial panel from an automated system, the MICs for each drug were determined according to the provided manufacturer's instructions and were as follows: meropenem, ≤0.25 to >2 µg/mL; imipenem, ≤0.5 to >2 µg/mL; ceftazidime, ≤1 to >8 µg/mL; cefoperazone/sulbactam, ≤8 to >32 µg/mL; cefmetazole, ≤4 to >32 µg/mL; ceftriaxone, ≤0.5 to >2 µg/mL; cefepime, ≤1 to >16 µg/mL; levofloxacin, ≤0.12 to >4 µg/mL; tazobactam/piperacillin, ≤4 to >64 µg/mL, and gentamicin, ≤2 to >8 µg/mL.

4.3. Definition and Breakpoint of Screening Antibiotics

The criteria for drug susceptibility were based on the CLSI guidelines, with MIC values of ceftazidime ≥8 µg/mL. However, the CLSI has not established breakpoints for cefoperazone/sulbactam in *Enterobacterales*. Therefore, referring to the breakpoint of cefoperazone in past studies, the cut-off MIC for cefoperazone/sulbactam was set to ≥32 µg/mL [11,28]. When a bacterial strain showed resistance to both agents, it was defined as a positive result. CRE isolates were identified based on criteria defined by the Japanese Ministry of Health, Labour and Welfare, which defines CRE as having an MIC ≥ 2 µg/mL for meropenem, MIC ≥ 2 µg/mL for imipenem, and MIC > 32 µg/mL for cefmetazole.

4.4. Identification of Metallo-Beta-Lactamase

The carbapenemase confirmation test was performed using the SMA test with an SMA Disk EIKEN® (Eiken Chemical Co. Ltd., Tokyo, Japan), according to the manufacturer's instructions. The tested bacteria were streaked onto Mueller–Hinton agar using a cotton swab after being inoculated and adjusted to the same turbidity as the McFarland standard of 0.5 in sterile saline solution. Two CAZ disks were placed on the agar, at least 3 cm apart from each other. An SMA disk was placed approximately 1.5–2 cm away from the center of one of the CAZ disks. Within 15 min, the plates were placed in a 35 °C incubator and cultured for 16–18 h. The inhibition zones were measured after incubation. If the inhibition zone of CAZ adjacent to the SMA disk expanded by ≥5 mm perpendicular to the axis connecting the centers of the SMA and CAZ disks, the tested bacteria were considered as likely producers of MBL. If no inhibition zone was observed using the SMA disk method, an imipenem disk was used instead. For all positive isolates in the SMA test and for all CRE isolates, subsequent PCR testing was performed. Carbapenemase genes in the CRE isolates were identified using a Cica Geneus® Carbapenemase Genotype Detection KIT2 (Kanto Chemical Co., Tokyo, Japan), which can detect the following genes: IMP-1, VIM, GES, KPC, NDM, OXA-48, and IMP-6. Bacterial processing and thermal cycling were performed according to the manufacturer's instructions.

4.5. Statistical Analysis

Fisher's exact test was used to compare the proportion between two independent groups. Sensitivity and specificity analyses were performed using common methods. The statistical significance of sensitivity and specificity was analyzed using R statistical software (version 4.1.2; R Foundation for Statistical Computing, Vienna, Austria) with a significance level of $p < 0.05$.

5. Conclusions

In conclusion, we demonstrated that screening with ceftazidime and cefoperazone/sulbactam, as non-carbapenem agents, can effectively detect MBL-producing *Enterobacterales* strains, including those susceptible to meropenem. The EUCAST method is a highly effective approach; however, this screening method showed comparable results to screening with meropenem and may serve as a supplementary approach, even for MBL strains with a meropenem MIC $\leq$ 0.25 µg/mL. Additionally, the use of ceftazidime and cefoperazone/sulbactam for screening showed utility in excluding non-MBL-producing CRE strains. Therefore, the ceftazidime and cefoperazone/sulbactam screening method can be effectively utilized, particularly in regions where IMP-type MBLs are predominant.

Supplementary Materials: The following supporting information can be downloaded at: https://www.mdpi.com/article/10.3390/antibiotics12071146/s1, Table S1: The 3776 non-duplicate sequential isolates. Table S2: Classification of CRE strains based on drug susceptibility to ceftazidime and cefoperazone/sulbactam.

Author Contributions: Conceptualization, K.T. (Kentarou Takei), H.K. and K.T (Koichi Tokuda); data curation and interpretation, K.T. (Kentarou Takei), H.K. and K.T. (Koichi Tokuda); inspection enforcement, M.C., A.N. and Y.T.; supervision, I.S., C.K., H.B., K.O., A.N., M.C., H.K. and K.T. (Koichi Tokuda); visualization, K.T. (Kentarou Takei); writing—original draft, K.T. (Kentarou Takei) All authors have read and agreed to the published version of the manuscript.

Funding: This work was supported by Japan Society for the Promotion of Science (JSPS) Grants-in-Aid for Scientific Research (KAKENHI) Grant Number 21K08504.

Institutional Review Board Statement: This study was approved by the Institutional Review Board of Tohoku University Graduate School of Medicine (IRB Numbers: 2018-1-445 and 2022-1-863).

Informed Consent Statement: Not applicable.

Data Availability Statement: The datasets of the current study are not publicly due to the fact that they contain a great deal of detailed patient's information. The dataset is owned by the Department of Infectious Diseases, Internal Medicine, Tohoku University Graduate School.

Acknowledgments: We are grateful to Masahiro Toyokawa (Department of Clinical Laboratory Sciences, School of Health Sciences, Fukushima Medical University) for his valuable feedback on the overall content of this paper.

Conflicts of Interest: The authors declare no conflict of interest.

References

1. Ambler, R.P. The structure of beta-lactamases. *Philos. Trans. R. Soc. Lond. B Biol. Sci.* **1980**, *289*, 321–331. [CrossRef] [PubMed]
2. Osano, E.; Arakawa, Y.; Wacharotayankun, R.; Ohta, M.; Horii, T.; Ito, H.; Yoshimura, F.; Kato, N. Molecular characterization of an enterobacterial metallo beta-lactamase found in a clinical isolate of Serratia marcescens that shows imipenem resistance. *Antimicrob. Agents Chemother.* **1994**, *38*, 71–78. [CrossRef]

3. Mori, N.; Tada, T.; Oshiro, S.; Kuwahara-Arai, K.; Kirikae, T.; Uehara, Y. A transferrable IncL/M plasmid harboring a gen encoding IMP-1 metallo-β-lactamase in clinical isolates of Enterobacteriaceae. *BMC Infect. Dis.* **2021**, *21*, 1061. [CrossRef] [PubMed]
4. Yigit, H.; Queenan, A.M.; Anderson, G.J.; Domenech-Sanchez, A.; Biddle, J.W.; Steward, C.D.; Alberti, S.; Bush, K.; Tenover, F.C Novel carbapenem-hydrolyzing β-lactamase, KPC-1, from a carbapenem-resistant strain of Klebsiella pneumoniae. *Antimicrob Agents Chemother.* **2001**, *45*, 1151–1161. [CrossRef] [PubMed]
5. Yong, D.; Toleman, M.A.; Giske, C.G.; Cho, H.S.; Sundman, K.; Lee, K.; Walsh, T.R. Characterization of a new metallo-beta lactamase gene, bla (NDM-1), and a novel erythromycin esterase gene carried on a unique genetic structure in Klebsiella pneumoniae sequence type 14 from India. *Antimicrob. Agents Chemother.* **2009**, *53*, 5046–5054. [CrossRef] [PubMed]
6. Centers for Disease Control and Prevention (Press Release). Action Needed Now to Halt Spread of Deadly Bacteria. Available online: https://www.cdc.gov/media/releases/2013/p0305_deadly_bacteria.html (accessed on 14 May 2023).
7. Centers for Disease Control and Prevention (CDC). Antibiotic Resistance Threats in the United States. 2019. Available online https://www.cdc.gov/drugresistance/pdf/threats-report/2019-ar-threatsreport-508.pdf (accessed on 8 April 2023).
8. Kamio, K.; Espinoza, J.L. The predominance of Klebsiella aerogenes among carbapenem-resistant Enterobacteriaceae infections in Japan. *Pathogens* **2022**, *11*, 722. [CrossRef] [PubMed]
9. European Committee on Antimicrobial Susceptibility Testing. EUCAST Guidelines for Detection of Resistance Mechanisms and Specific Resistances of Clinical and/or Epidemiological Importance. 2017. Available online: http://www.eucast.org/resistance_mechanisms/ (accessed on 20 March 2023).
10. Clinical and Laboratory Standards Institute (CLSI). *Performance Standards for Antimicrobial Susceptibility Testing, CLSI Supplement M100-S30*; CLSI: Wayne, PA, USA, 2020.
11. Nishio, H.; Komatsu, M.; Shibata, N.; Shimakawa, K.; Sueyoshi, N.; Ura, T.; Satoh, K.; Toyokawa, M.; Nakamura, T.; Wada, Y et al. Metallo-beta-lactamase-producing gram-negative bacilli: Laboratory-based surveillance in cooperation with 13 clinical laboratories in the Kinki region of Japan. *J. Clin. Microbiol.* **2004**, *42*, 5256–5263. [CrossRef]
12. Fattouh, R.; Tijet, N.; McGeer, A.; Poutanen, S.M.; Melano, R.G.; Patel, S.N. What is the appropriate meropenem MIC for screening of carbapenemase-producing Enterobacteriaceae in low-prevalence settings? *Antimicrob. Agents Chemother.* **2016**, *60*, 1556–1559 [CrossRef]
13. Richter, S.; Marchaim, D. Screening for carbapenem-resistant Enterobacteriaceae: Who, when, and how? *Virulence* **2017**, *8*, 417–426 [CrossRef]
14. Japanese Society of Chemotherapy; Japanese Society of Infectious Disease; Japanese Society for Infection Prevention and Control Japanese Society for Clinical Microbiology. Joint Statement for Management of Enterobacteriaceae with Lower Susceptibility to Carbapenems: Importance of Carbapenemase-Producing Bacteria as Infection Control Targets [Japanese]. Available online https://www.chemotherapy.or.jp/uploads/files/guideline/4gakkai2017_b.pdf (accessed on 20 March 2023).
15. Pranita, D.T.; Yohei, D.; Robert, A.B.; JKristie, J.; Patricia, J.S. A Primer on AmpC β-Lactamases: Necessary Knowledge for an Increasingly Multidrug-resistant World. *Clin. Infect. Dis.* **2019**, *69*, 1446–1455. [CrossRef]
16. Ma, D.Y.; Huang, H.Y.; Zou, H.; Wu, M.L.; Lin, Q.X.; Liu, B.; Huang, S.F. Carbapenem-resistant *Klebsiella aerogenes* clinical isolates from a teaching hospital in southwestern China: Detailed molecular epidemiology, resistance determinants, risk factors and clinical outcomes. *Infect. Drug Resist.* **2020**, *19*, 577–585. [CrossRef]
17. Annavajhala, M.K.; Gomez-Simmonds, A.; Uhlemann, A.-C. Multidrug-resistant *Enterobacter cloacae* complex emerging as a global, diversifying threat. *Front. Microbiol.* **2019**, *10*, 44. [CrossRef]
18. Lee, J.-Y.; Hong, Y.-K.; Lee, H.; Ko, K.S. High prevalence of non-clonal imipenem-nonsusceptible *Enterobacter* spp. isolates in Korea and their association with porin down-regulation. *Diagn. Microbiol. Infect. Dis.* **2017**, *87*, 53–59. [CrossRef]
19. Idrees, M.M.; Rimsha, R.; Idrees, S.A. Antimicrobial susceptibility and genetic prevalence of extended-spectrum β-lactamases in gram-negative rods isolated from clinical specimens in Pakistan. *Antibiotics* **2022**, *12*, 29. [CrossRef]
20. Lin, S.Y.; Lu, P.L.; Wu, T.S.; Shie, S.-S.; Chang, F.-Y.; Yang, Y.-S.; Chiang, T.-T.; Wang, F.-D.; Ho, M.-W.; Chou, C.-H.; et al. Correlation between cefoperazone/sulbactam MIC values and clinical outcomes of Escherichia coli bacteremia. *Infect. Dis. Ther.* **2022**, *11*, 1853–1867. [CrossRef]
21. Hassoun-Kheir, N.; Hussein, K.; Karram, M.; Saffuri, M.; Badaan, S.; Peleg, S.; de Kraker, M.E.; Aboelhega, W.; Warman, S. Alon, T.; et al. Risk factors for acquisition of carbapenemase-producing versus non-carbapenemase-producing *Enterobacterales*: A case-control study. *Clin. Microbiol. Infect.* **2023**, *29*, 629–634. [CrossRef]
22. Jong, H.L.; Il, K.B.; Chae, H.L.; Seri, J. Molecular characteristics of first IMP-4-producing *Enterobacter cloacae* sequence type 74 and 194 in Korea. *Front. Microbiol.* **2017**, *8*, 2343. [CrossRef]
23. Latania, K.L.; Robert, A.W. The epidemiology of carbapenem-resistant Enterobacteriaceae: The impact and evolution of a global menace. *J. Infect. Dis.* **2017**, *215*, S28–S36. [CrossRef]
24. Arakawa, Y. Systematic research to overcome newly emerged multidrug-resistant bacteria. *Microbiol. Immunol.* **2020**, *64*, 231–251 [CrossRef]
25. Ronat, J.B.; Oueslati, S.; Natale, A.; Kesteman, T.; Elamin, W.; Langendorf, C.; Hardy, L.; Vandenberg, O.; Naas, T. Validation of Three MicroScan® Antimicrobial Susceptibility Testing Plates Designed for Low-Resource Settings. *Diagnostics* **2022**, *12*, 2106 [CrossRef]

6. Humphries, R.M.; Ambler, J.; Mitchell, S.L.; Castanheira, M.; Dingle, T.; Hindler, J.A.; Koeth, L.; Seii, K.; on behalf of the CLSI Methods Development and Standardization Working Group of the Subcommittee on Antimicrobial Susceptibility Testing. CLSI Methods Development and Standardization Working Group best practices for evaluation of antimicrobial susceptibility tests. *J. Clin. Microbiol.* **2018**, *56*, e01934-17. [CrossRef] [PubMed]
7. Clinical and Laboratory Standards Institute (CLSI). Implementation of the CLSI Method for Direct Disk Diffusion Testing from Positive Blood Culture. Available online: https://clsi.org/about/blog/implementation-of-the-clsi-method-for-direct-disk-diffusion-testing-from-positive-blood-cultures/ (accessed on 12 June 2023).
8. Sader, H.S.; Carvalhaes, C.G.; Streit, J.M.; Castanheira, M.; Flamm, R.K. Antimicrobial activity of cefoperazone-sulbactam tested against Gram-Negative organisms from Europe, Asia-Pacific, and Latin America. *Int. J. Infect Dis.* **2020**, *91*, 32–37. [CrossRef] [PubMed]

antibiotics

Article

Antimicrobial Potential of Metabolites in Fungal Strains Isolated from a Polluted Stream: *Annulohypoxylon stygium* WL1B5 Produces Metabolites against Extended-Spectrum Beta-Lactamase-Positive *Escherichia coli*

Walter Oliva Pinto Filho Segundo [1], Roberta Silva de Oliveira [2], Rildo Mendes Lima [3], Paulo Alexandre Lima Santiago [4], Luciana Aires de Oliveira [1], Ana Cláudia Alves Cortez [2], Emerson Silva Lima [5], Érica Simplício de Souza [6], Hagen Frickmann [7,8,*] and João Vicente Braga de Souza [2,*]

1 Program in Biodiversity and Biotechnology of the Bionorte Network (PPG-BIONORTE), Amazonas State University (UEA), Manaus 69065-001, AM, Brazil
2 Mycology Laboratory, National Institute for Amazonian Research (INPA), Manaus 69067-375, AM, Brazil
3 Central Public Health Laboratory of the State of Amazonas (LACEN/AM), Manaus 69020-040, AM, Brazil
4 Department of Biology, Amazonas State University (UEA), Tabatinga 69640-000, AM, Brazil
5 Faculty of Pharmaceutical Sciences, Federal University of Amazonas (UFAM), Manaus 69067-005, AM, Brazil
6 Higher School of Technology, Amazonas State University (UEA), Manaus 69050-020, AM, Brazil
7 Department of Medical Microbiology, Virology and Hygiene, University Medicine Rostock, 18057 Rostock, Germany
8 Department of Microbiology and Hospital Hygiene, Bundeswehr Hospital Hamburg, 20359 Hamburg, Germany
* Correspondence: frickmann@bnitm.de (H.F.); joaovicentebragasouza@yahoo.com.br (J.V.B.d.S.)

Abstract: The emergence of multidrug resistance in bacterial pathogens is a growing public health concern requiring solutions including the discovery of new antimicrobial drugs. Fungi have been used for decades as a source of antimicrobials. Ongoing screenings for newly characterized fungal strains producing antimicrobials include environments that are difficult to access like the deep sea, glaciers, wastewaters and environments polluted due to human activity. In the present study, fungal microorganisms were isolated from water samples taken from a polluted stream in the city of Manaus, AM, Brazil, and screened for antimicrobial effects against *Escherichia coli*. Using extracts from five isolates (*Annulohypoxylon stygium* WL1B5, *Colletotrichum fructicola* WL3B9, *Clonostachys rosea* WL5B18, *Clonostachys rosea* WL8B28 and *Trichoderma harzianum* WL9B49), antimicrobial activity against the reference strains *Escherichia coli* ATCC 25922 as well as *E. coli* NCTC 13353, an extended-spectrum beta-lactamase-positive strain, was observed. Inhibition zones ranged from 1 to 35.9 mm and a minimum inhibitory concentration of 400 μg/mL could be demonstrated. Assessments of the metabolites of *Annulohypoxylon stygium* WL1B5 allowed us to identify nodulisporone and daidzein, which have already been associated with antimicrobial activity. The findings confirm the feasibility of isolating fungal strains from polluted sites producing metabolites that can serve as potential future alternatives for the treatment of multidrug-resistant bacteria.

Keywords: fungi; antimicrobial; bacteria; resistance; *Escherichia coli*

1. Introduction

The international emergence of multidrug-resistant or even pan-drug-resistant bacterial pathogens is an ongoing concern of global healthcare and public health. The World Health Organization (WHO) predicts that by 2050, multidrug-resistant bacteria will likely be causally associated with the death of approximately 10 million people per year [1].

Starting with the clinical application of antibiotic drugs (e.g., penicillins and cephalosporines) several decades ago, evolutionary selection in many bacterial species has driven the spread of various distinct mechanisms of resistance to several classes of antibiotics used for therapeutic purposes in both outpatient medical facilities and hospitals [2]. Bacterial pathogens such as methicillin-resistant *Staphylococcus aureus* (MRSA), vancomycin-resistant *Enterococcus* spp. (VRE), penicillin-resistant *Streptococcus pneumoniae* and extended-spectrum beta-lactamase (ESBL)-producing Enterobacterales have emerged, posing a challenge to antimicrobial therapies [3].

Filamentous fungi from the phylum Ascomycota are known to produce metabolites with antimicrobial activity such as alkaloids, terpenoids, polyketides, polysaccharides, steroids, quinones, flavonoids, aliphatic compounds and phenols [4]. As recently summarized, most new, naturally occurring substances with antimicrobial activity described in the literature have been isolated from fungi; several such compounds have facilitated the development of new, clinically important antimicrobial agents [5]. As bacterial resistance is a constant concern, there is an ongoing need for new antimicrobial substances with sufficient biological activity and low therapeutic side effects allowing their use for medical therapies.

Severely polluted streams have been shown to harbor a high number of fungal species with the potential of producing bioactive metabolites [6,7]. Associated with the strong selection pressure on local microbial sentinel communities in such toxic environments, fungi isolated from such locations have been shown to produce more potent secondary metabolites than those that are commonly produced under stabler conditions. Therefore, such environments are likely to host a wide variety of fungal strains with promising antimicrobial bioactivity. As a proof-of-principle assessment, the objective of this work was to investigate the presence of cultivable filamentous fungi in a severely polluted stream in Manaus City, Amazonas, Brazil, and their ability to produce substances with antimicrobial activity.

2. Results

2.1. Collection Sites

Physical and chemical analyses were carried out with water taken at various sampling sites on the Mindú stream (Table 1, Appendix A Figure A1). The data indicate that the sampling points "central point" and "estuary" were particularly impacted by pollution due to industrial or domestic waste being dumped into the water. Even though the temperature and pH of the water remained similar, the changes of other variables such as dissolved oxygen, turbidity, electrical conductivity and total nitrogen indicated inconsistent water quality between the different sampling sites.

Table 1. Mean values and standard deviation of the limnological parameters measured at the various sampling sites.

Limnological Parameters	Sampling Sites at Mindú Stream			
	Unit	Head Water	Central Point	Estuary
Temperature	°C	27 ± 0.1	27 ± 0.1	27 ± 0.1
pH	-	6.3 ± 0.1	7. 3 ± 0.2	6.1 ± 0.1
Dissolved oxygen (DO)	(Sat) %	3.2 ± 0.1	0.1 ± 0.1	0.1 ± 0.1
Chemical oxygen demand O_2 (Conc.)	mgO_2/L	57 ± 0.1	107 ± 0.2	151 ± 0.3
Turbidity	NTU	3.4 ± 0.2	3.9 ± 0.2	10 ± 0.3
Eletric conductivity	μScm^{-1}	0.4 ± 0.1	47.6 ± 0.4	106.8 ± 0.1
Nitrate	mg/L	0.1 ± 0.1	0.2 ± 0.1	2 ± 0.1
Nitrite	mg/L	0.1 ± 0.1	0.1 ± 0.1	0.2 ± 0.1
Amonia	mg/L	0.1 ± 0.1	0.2 ± 0.0	1.2 ± 0.1
Total nitrogen	mg/L	1.4 ± 0.1	2.4 ± 0.6	8.7 ± 0.1
Total phosphorus	mg/L	0.1 ± 0.1	0.2 ± 0.2	0.4 ± 0.1
Orthophosphate	mg/L	0.1 ± 0.1	0.1 ± 0.1	0.3 ± 0.1

2.2. Fungal Isolates

A total of 67 fungal isolates were obtained from the sampling sites. Conventional identification based on morphological assessments allowed the taxonomic assignment of the genera of 21 isolates (Table 2). In addition, 5 isolates that were identified as antimicrobial substance producers in the present work were further differentiated at the species level—using DNA sequencing based on ITS1-5.8S rRNA gene-ITS2 and 26 rRNA gene fragments—as *Annulohypoxylon stygium* (sample ID WL1B5), *Colletotrichum fructicola* (samples ID WL3B9), *Clonostachys rosea* (sample ID WL5B18) (syn. *Gliocladium roseum*), *Clonostachys rosea* (sample ID WL8B28) and *Trichoderma harzianum* (sample ID WL9B49) (Figure 1). The nucleotide sequences obtained after amplification and sequencing were deposited at NCBI GenBank with the accession numbers OP580168, OP578038, OP563025, OP578133 and OP578134, respectively.

Table 2. Number of CFU (colony forming units)/mL, the total number of fungal isolates and the total number of isolates of aerobic fungi from three different sampling sites on the Mindú stream, Manaus, Amazonas. CFU = colony forming units.

Sample Site	CFU/mL	Total Isolates	Genera	Number of Isolates by Genus
Head water (03 01′07.31″ S 59 55′84″ W)	1.4 × 10	14	*Colletotrichum* sp. *Oidiodendrum* sp. *Mycelia sterilia*	1 1 12
Central point (03 04′28.4″ S 59 58′57.7″ W)	2.4 × 10	24	*Annulohypoxylon* sp. *Chloridium* sp. *Clonostachys* sp. *Oidiodendrum* sp. *Paecilomyces* sp. *Rodothorula* sp. *Trichoderma* sp. *Mycelia sterillia*	1 1 1 1 1 1 1 17
Estuary (03 06′58.4″ S 60 02′01.5″ W)	2.9 × 10	29	*Aspergillus* sp. *Clonostachys* sp. *Fusarium* sp. *Penicillium* sp. *Rodothorula* sp. *Trichoderma* sp. *Mycelia sterillia*	1 1 5 1 1 4 17

2.3. Production of Antibacterial Substances by the Fungal Isolates

In order to investigate the potential production of substances with antibacterial activity by the fungal isolates, all 67 isolates were subjected to submerged bioprocessing for 14 days. The fungal cultures were extracted applying a hexane- and EtOAc-based procedure, as detailed in the methods chapter. The obtained fractions were tested by agar diffusion testing for antimicrobial activity against *E. coli* ATCC 25922 and the extended-spectrum beta-lactamase- (EBSL-)producing strain *E. coli* NCTC 13353. Table 3 shows the antibacterial activity of hexane-based, EtOAc-based and crude extracts with inhibition zone diameters ranging from 1 to 35.9 mm. Among the 67 fungal isolates investigated, metabolites of 5 isolates were associated with halos of bacterial growth inhibition. Metabolites from the isolates *A. stygium* (WL1B5), *C. rosea* (WL5B18) and *C. rosea* (WL8B28) were found to induce the largest inhibition halos when tested with *E. coli* ATCC 25922 and the ESBL-positive *E. coli* strain NCTC 13353.

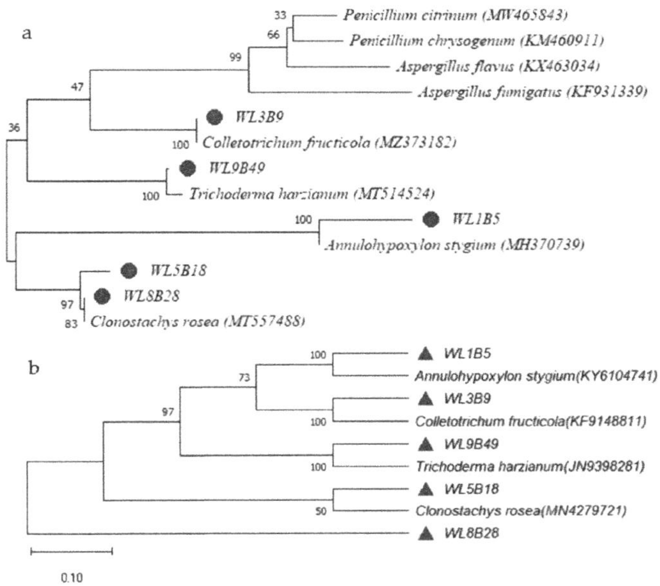

Figure 1. Phylogenetic tree constructed based on the sequences of ITS1-5.8S rRNA gene-ITS2 (**a**) and the D1/D2 domain of the 28S rRNA gene (**b**) using the maximum likelihood method and the Tamura-Nei model [8]. The percentage of trees in which the associated taxa clustered together in the bootstrap test (1000 replicates) is shown next to the branches. The tree is drawn to scale, with branch lengths indicating the number of substitutions per site. This analysis involved 22 nucleotide sequences. The green dots/pyramids indicate the isolates from this study. The others are previously identified sequences of antimicrobial substance-producing fungi. There were a total of 1318 positions in the final dataset. Evolutionary analyses were conducted by applying MEGA X v10.2.4.

Table 3. Antibacterial activity of different solvent extracts based on hexane, EtOAc and crude extraction of fungal isolates from the Mindú stream assessed with *E. coli* ATCC 25922 and an ESBL-positive *E. coli* NCTC 13353.

Fungal Species	Zones of Inhibition (mm)					
	Bacterial Strains					
	E. coli ATCC 25922			*E. coli* NCTC 13353		
	Hex	EtOAc	CE	Hex	EtOAc	CE
Annulohypoxylon stygium WL1B5	-	35.9 ± 0.1	-	-	14 ± 1.4	5.95 ± 0.1
Colletotrichum fructicola WL3B9	7.9 ± 0.1	7.25 ± 0.4	6 mm	-	-	-
Clonostachys rosea WL5B18	-	19.8 ± 0.3	13.9 ± 0.1	-	4 ± 1.4	-
Clonostachys rosea WL8B28	1 ± 1.4	13.15 ± 1.2	26.95 ± 0.1	-	8 ± 1.4	18.9 ± 0.1
Trichoderma harzianum WL9B49	-	4.4 ± 0.8	-	-	-	-
Positive control (amoxillin 50 µg/mL)		20 mm			0 mm	
Negative control (10% DMSO)		0 mm			0 mm	

Hex—Hexane extract; EtOAc—Ethyl acetate extract; CE—Crude extract; extract volume 100 µL; incubation time and temperature 24 h/37 °C. The dash symbol (-) indicates no activity.

Minimum Inhibitory Concentration—MIC

The minimal inhibitory concentrations of EtOAc extracts of *A. stygium* WL1B5, *C. rosea* WL5B18 and *C. rosea* WL8B28 were assessed by applying a microdilution assay with *E. coli* ATCC 25922 and the ESBL-positive strain *E. coli* NCTC 13353. The EtOAc extracts of these three fungal isolates yielded a MIC of 400 μg/mL for both assessed bacterial strains (Table 4).

Table 4. Minimum inhibitory concentrations (MIC, μg/mL) of extracts of selected fungi with antimicrobial activity isolated from the Mindú stream.

Extract	Minimum Inhibitory Concentration of EtOAc Extracts (μg/mL)		
	Fungal Species and Strain Identity	Bacterial Strains	
		E. coli ATCC 25922	*E. coli* NCTC 13353
EtOAc	*Annulohypoxylon stygium* WL1B5	400 μg/mL	400 μg/mL
	Clonostachys rosea WL5B18	400 μg/mL	400 μg/mL
	Clonostachys rosea WL8B28	400 μg/mL	400 μg/mL
Positive control		amoxillin (12.5 μg/mL)	
Negative control		10% DMSO	

Initial extract concentration: 3.2 mg/mL; incubation time and temperature: 16/20 h/37 °C.

2.4. In Vitro Cytotoxicity Assay

In order to study the toxicity of the fungal metabolites, the cytotoxicity, i.e., lethal effect on cell viability, of the metabolites of EtOAc extracts of *A. stygium* WL1B5, *C. rosea* WL5B18 and *C. rosea* WL8B28 was assessed with MRC5 human lung fibroblast cells. The EtOAc extracts of *A. stygium* WL1B5, *C. rosea* WL5B18 and *C. rosea* WL8B28 showed an IC50% > 100 μg/mL, indicating low toxicity to human cells of the MRC5 line. Doxorubicin, which was used as an experimental positive control, showed a much lower IC50% of 5 μg/mL.

2.5. TLC and Contact Bioautography

TLC assays for the analysis of EtOAc extracts were carried out with *A. stygium* WL1B5, *C. rosea* WL5B18 and *C. rosea* WL8B28 using the mobile phases hexane (100%), hexane/ethyl acetate (1:1, *v/v*), ethyl acetate (100%), ethyl acetate/methanol (1:1, *v/v*) and methanol (100%). In addition, a bioautography assay was conducted with the TLC platelets to identify the chromatographic bands with antibacterial activity. Table 5 presents the mobile phase, main bands and the Rf value of the bands that were associated with antibacterial activity against *E. coli* ATCC 25922 (Appendix A Figure A2).

Table 5. TLC analysis and contact bioautography of mobile phases tested against *E. coli* ATCC 25922 with Rf values.

EtOAc Extract	Mobile Phase	Rf of Bands with Antimicrobial Activity
A. stygium WL1B5	Hex (100%)	0.15
	MeOH (100%)	0.86
C. rosea WL5B18	EtOAc/MeOH (1:1)	0.71
C. rosea WL8B28	EtOAc/MeOH (1:1)	0.78

Rf—retention factor; Hex—hexane; MeOH—methanol; EtOAc/MeOH—ethyl acetate/methanol.

2.6. LC-MS/MS Analysis

When the EtOAc extract of *A. stygium* WL1B5 was further investigated using LC-MS/MS, the chromatograms showed three major peaks at 3.8, 4.7 and 7.3 min (Appendix A Figure A3). The peak eluting at 3.8 min had the main ion compounds of [M-H] *m/z* 193 and [M + H] *m/z* 212, the peak eluting at 4.7 had no ion compounds equivalent to the mass found and the peak eluting at 7.3 min had the main ion compounds of [M-H] *m/z* 234 and [M + H] *m/z* 208 (Appendix A Figures A4 and A5). Figure 2 presents likely metabolites abundant in the EtOAc extract of *A. stygium* WL1B5. Two substances were suggested by the LC-MS/MS results: nodulisporone sodium (**1**) and daidzein (**2**).

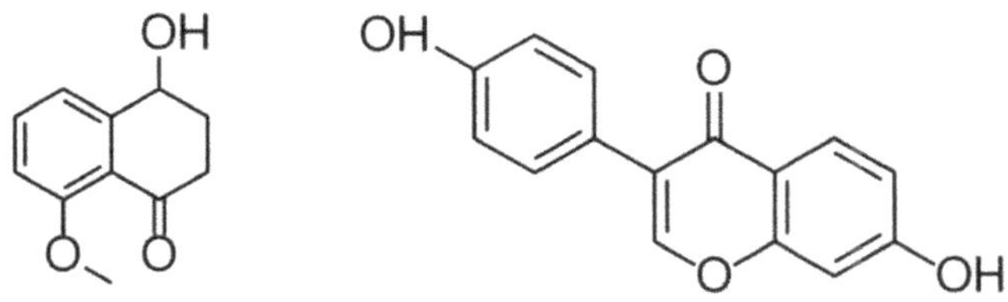

1 2

Figure 2. Structures of nodulisporone sodium (**1**) and daidzein (**2**) detected by LC-MS/MS in the EtOAc extract of the fungus *A. stygium* WL1B5. Chemical structures were drawn using the ChemDraw Professional 18.1 software (PerkinElmer Informatics, Waltham, Massachusetts, United States).

3. Discussion

The findings of this study confirm that isolated fungal strains from polluted environments like streams show the potential to produce antimicrobially active substances. In summary, *A. stygium* WL1B5, *C. fructicola* WL3B9, *C. rosea* WL5B18, *C. rosea* WL8B28 and *T. harzianum* WL9B49 were able to synthesize metabolites with activity against *E. coli* ATCC 25922 and the ESBL-positive *E. coli* strain NCTC 13353 when grown under submerged fermentation conditions. Applying EtOAc extraction, the most pronounced antimicrobial activity could be demonstrated. With the help of LC-MS/MS analysis, two active metabolites were proposed in the EtOAc extract of the *A. stygium* WL1B5 isolate.

The assessed water samples indicated an increase in pollution between the sampling sites "head waters" and "estuary", suggesting a continuous load of domestic/industrial waste which is dumped in the Mindú stream (Appendix A Figure A1) [9]. Concomitantly, there was an increase in the number of isolated fungi. Most of the fungi were plant-associated, and a relevant proportion could not be identified due to the absence of reproductive elements. This result was expected, because in addition to domestic sewage, the Mindú stream also harbors riparian vegetation, which is characteristic of the Amazon region.

Our findings further demonstrated that the isolates *Annulohypoxylon stygium* WL1B5, *Colletotrichum fructicola* WL3B9, *Clonostachys rosea* WL5B18, *Clonostachys rosea* WL8B28 and *Trichoderma harzianum* WL9B49 produced metabolites with antibacterial activity. Specifically, metabolites in the EtOAc extracts of *Annulohypoxylon stygium* WL1B5, *Clonostachys rosea* WL5B18, *Clonostachys rosea* WL8B28 could be associated with activity against an *E. coli* reference strain with susceptibility towards third-generation cephalosporines and the ESBL-positive strain *E. coli* NCTC 13353. This finding matches with previous works that demonstrated activity of metabolites of *A. stygium* against *Bacillus subtilis*, *E. coli* ATCC 25922, methicillin-resistant *S. aureus* (MRSA), *S. aureus* ATCC 25923 and *Candida albicans* [10–13]. The results are also in line with previous works carried out with *C. rosea* species, which had suggested antimicrobial activity of fungal metabolites against *E. coli*, *S. aureus*, *Bacillus cereus* and *Citrobacter freundii* [14–17].

Of note, the AcOEt extracts of *A. stygium* WL1B5, *C. rosea* WL5B18 and *C. rosea* WL8B28 were shown to be associated with only a minor cytotoxic effect on MRC5 cells. This finding is in agreement with previously published results on the comparatively low toxicity of *A. stygium* metabolites [17,18]. Similar results have also been reported for fungal metabolites produced by *Clonostachys rosea* [19,20].

HLPC-MS/MS analysis of the EtOAc extract of *A. stygium* WL1B5 proposed the abundance of the metabolites nodulisporone and daidzein. Nodulisporone is a tetralone originally isolated from fungi of the *Nodulisporium* genus (Ascomycota, *Xylariacea*) [21] and recently also demonstrated in fungi of the *Annulohypoxylon* genus [22]. In EtOAc extracts of *Xylaria* polymorpha (*Xilariaceae*), this compound has previously shown antimicrobial inhibition zones with *Candida albicans* strains (10 mm) and *T. harzianum* strains [13 mm) [22]. Tetralones have been described to interfere with DNA replication by inhibiting the cellular DNA topoisomerase, thus showing effects on gene expression and recombination of microorganisms [23].

Daidzein is a well-characterized flavonoid produced by plants with several pharmacological properties. This metabolite has already been identified in the genera *Aspergillus*, *Penicillium* and *Trichoderma* [24–26]. However, this is the first time that daidzein is reported in an *Annuloxypoxylon* species. When previously isolated from *Trichoderma* sp. YM 311505, this compound had exhibited antibacterial activity against *E. coli* with a minimum inhibitory concentration (MIC) of 64 µg/mL [26]. As a component of of EtOAc extracts of *Aspergillus tubingensis*, antimicrobial activity against several strains of *Vibrio* species with MIC values between 0.5 µg/mL to 1.0 µg/mL have been associated with this substance [27]. Flavonoids are known for their inhibiting effects on bacterial nucleic acid synthesis and metabolism as well as for effects on the cellular membrane permeability [28].

The present work opens perspectives for future studies focusing on a) a deeper identification of bioactive molecules responsible for the observed antibacterial effects; b) test assays for antimicrobial effects including Gram-positive strains and c) elucidation of the substances applying nuclear magnetic resonance (NMR) analyses. Hence, the here presented data can serve as a basis for ongoing studies, contributing to a better understanding of the chemical and biological characteristics of the isolates. The discovery of new bioactive substances such as natural antibiotics as important resources to counteract the problem of increasing drug resistance in Gram-negative bacterial pathogens like ESBL-positive *E. coli* will remain a relevant focus for future research activities.

4. Materials and Methods

4.1. Sample Collection

Five water samples were collected from the polluted stream Mindú at three sampling sites in Manaus city, Amazonas, Brazil. In the following, these sampling sites are addressed as "head water" (coordinates 03 01′07.31″ S 59 55′84″ W), "central point" (coordinates 03 04′28.4″ S 59 58′57.7″ W) and "estuary" coordinates (03 06′58.4″ S 60 02′01.5″ W). In line with the recommendations by the American Public Health Association Standards [29], all samples were taken from a depth of 15 to 30 cm, kept in sterile flasks, stored in an icebox at 4–8 °C and transported to the mycology laboratory of the National Institute of Amazonian Research for the isolation of fungi. Physical and chemical water parameters were measured on site using a Profline 197i Multiparameter Water Quality device (WTW, Wellheim, Germany).

4.2. Isolation of Fungi from the Polluted Water

Next, 1000 µL of sample material was inoculated via the spread plate technique on laboratory-produced Sabouraud agar (peptone 10 g/L, dextrose 20 g/L, agar 20 g/L) (Kasvi, Madrid, Spain) plates, to which the antibiotic drug chloramphenicol (250 mg/L) was added. Individual fungal strains were isolated by plating onto fresh Potato Dextrose Agar (PDA, Kasvi, Madrid, Spain) and incubating at 28 °C for 7 days. Each strain was identified to the genus level by assessing the morphological characteristics based on the microscopic description of conidia, phialides, metule, conidiophore and hyphae.

4.3. Molecular Fungal Identification

The selected fungi were identified based on both morphological and ITS (internal transcribed spacer) as well as 28S rRNA gene sequence analysis. Genomic DNA was isolated according to a previously described protocol [30]. The ITS sequence of the fungal isolate was amplified using the forward primer ITS1 (5′ CTTGGTCATTTAGAGGAAGTAA-3′) and the reverse primer ITS4 (5′ CTTGGTCATTTAGAGGAAGTAA-3′) as described elsewhere [31]. The D1/D2 domain of the 28S rRNA gene was amplified applying the primer pair NL1 (5′-GCATATCAATAAGCGGAGGAAAAG-3′) and NL4 (5′-GGTCCGTGTTTCAAGACGG-3′) using a published protocol [32]. Sequences of the fungal ITS and rRNA gene regions were compared with those deposited in the NCBI (National Center for Biotechnology Information; http://www.ncbi.nlm.nih.gov, last accessed on 19 October 2022) database. Multiple alignments of the ITS gene region were calculated using the software MUSCLE (default conditions for gap opening and gap extension penalties applied) and implemented in the software MEGA X version 10.2.4 (Molecular Evolutionary Genetics Analysis) [33]. The phylogenetic tree was calculated based on neighbor-joining (NJ) with bootstrap analysis including 1000 replicates.

4.4. Fungal Cultivation to Assess Secondary Metabolite Production

The fungal isolates were cultivated in Petri dishes containing PDA medium at room temperature for 7 to 10 days. The bioprocesses of fungal culture were performed as reported elsewhere [34] with minor modifications in triplicates. In short, the bioprocesses were carried out in an Erlenmeyer flask (150 mL) containing 50 mL of PDA broth (120 g/L potato, 10 g/L dextrose). About 1×10^4 fungal spores/mL were inoculated into this medium and incubated under static conditions allowing fermentation for 15 days in the dark at room temperature (28 °C ± 2 °C).

4.5. Preparation of Extracts

Subsequently, the obtained filtrates were extracted in order of polarity using the solvents hexane (Hex) and ethyl acetate (AcOEt) (Synth, São Paulo, Brazil) (1:1:3) to obtain the metabolites. The resulting broth extracts were dried in a vacuum evaporator and the residues were resuspended with 10% dimethyl sulfoxide (DMSO). All fungal extracts were evaluated for their biological/antibacterial activities.

4.6. Assessment of Antimicrobial Activity of Fungal Metabolites on E. coli Strains Applying the Well Plate Method

The antibacterial potential of fungal metabolites against bacterial isolates was tested applying the well diffusion method [35] on Mueller-Hinton agar (MHA, Kasvi, Madrid, Spain) plates. In short, 100 µL volumes of an *E. coli* ATCC (25922) cell suspension and a cell suspension of the extended-spectrum beta-lactamase-(ESBL-) positive strain *E. coli* NCTC® 13353 (Controllab, Rio de Janeiro, Brazil) with cell densities of 0.5 McFarland standard units each were inoculated on MH agar plates. Wells (6 mm) were loaded with 100 µL of extract each. Amoxillin at a concentration of 50 µg/mL was used as a positive control and 10% DMSO as a negative control. The plates were incubated for 24 h at 37 °C and the inhibition zone was recorded by measuring the diameter around the wells. Each experiment was performed in triplicates.

Minimum Inhibitory Concentration (MIC)

The minimum inhibitory concentration (MIC) was determined applying the modified microdilution method [36] with the same bacteria used for the diffusion approach described above. The fungal extracts were adjusted to an initial concentration of 3.2 mg/mL. The microdilution assay was performed in 96-well microtiter plates according to a previously described protocol. Briefly, fungal extracts were used in decreasing concentrations from 1.600 to 1.56 µg/mL. A volume of 10 µL bacterial inoculum was added with a final concentration of 10^5 CFU/mL. Incubation was carried out for 16 h to 20 h at 37 °C. The assays were performed in duplicate and the reading was performed after the incubation period based on visual comparison with bacterial growth in the positive control well. The MIC value was defined as the lowest concentration, expressed in µg/mL, at which microorganisms did not show visible growth.

4.7. Cytotoxicity Bioassays

Cytotoxicity of the extracts was evaluated applying the Alamar Blue™ assay [37] (Merck, Darmstadt, Germany) with MRC-5 (human fibroblast) cells grown in DMEM (Dulbecco's Medium Eagle's medium) supplemented with 10% fetal bovine serum kept in a CO_2 incubator at 37 °C with an atmosphere containing 5% CO_2. Cells were seeded in 96-well plates (0.5×10^4 cells per well). After 24 h, extracts were dissolved in DMSO, added to each well (10 µg/mL) and incubated for 48 h. Doxorubicin (5 µg m/L) was used as a positive control. Negative controls (blanks) received the same amount of DMSO and had the same final DMSO concentrations as the samples (0.1%). Two hours before the end of the incubation, 10 µL of Alamar Blue™ was added to each well. The resulting fluorescent signal was monitored with a multiplate reader at a range of 530–560 nm excitation wavelength and 590 nm emission wavelength. The fluorescent signals obtained with the assay were proportional to the number of vital cells in the sample.

4.8. Thin Layer Chromatography (TLC)

The selected extracts were dissolved in acetate (1 mg/mL) and applied with a fine-hole glass capillary tube. In detail, a volume of 10 µL of each fraction was applied over the entire length of the pre-mixed glass TLC plates coated with a silica gel 60 layer F254 (thickness 0.25 mm, Kieselgel 60 F254, Merck, Germany, size 20 × 20 cm). TLC chromatograms were obtained using optimized solvent systems (100% hexane, hexane/ethyl acetate 1:1—v/v, 100% ethyl acetate, ethyl acetate/methanol 1:1—v/v, 100% methanol) and examined under ultraviolet light at a wavelength of 365 nm. The metabolites were labeled.

4.9. Contact Bioautography

Contact bioautography was performed to determine the zone with antimicrobial activity as described [38] with minor modifications. The sterile materials obtained with the TLC approach described above were aseptically inoculated on the plates together with *E. coli* suspensions (MacFarland scale 0.5) and kept there for a period of 45 min in order to allow the metabolites to diffuse into the plates. Subsequently, the TLC platelets were removed from the surface of the plates with the help of sterile forceps and the plates were incubated in an inverted position for 48 h at 37 °C. Areas of inhibition were marked and the relevant Rf (retention factor) values were recorded and compared with the TLC chromatograms. All tests were performed in triplicates.

4.10. Liquid Chromatography-Mass Spectrometry (LC-MS/MS)

The crude extracts were analyzed with liquid chromatography coupled to mass spectrometry (Prominence UFLC, Shimadzu, Kyoto, Japan) and coupled to a mass spectrometer with quadrupole (Bruker Daltonics, Amazon Speed Ion Trap, Bremen, Germany). Five microliters of the dye samples were injected in C18 Shim-packs CLC-ODS(M)® (250 mm x 4.6 mm, particle diameter 5 µm) using a binary phase (solvent A: water, solvent B: methanol) for the separations. Gradient elution was conducted at concentrations of 20–100% solvent B over 24 min at a flow rate of 1 mL/min. The obtained spectra were assessed based on visual interpretation of MS/MS spectral data.

5. Conclusions

The present study investigated the antimicrobial potential of metabolites of fungi isolated from a polluted stream in Manaus, Amazonas, Brazil. Five isolates (*A. stygium* WL1B5, *C. fructicola* WL3B9, *C. rosea* WL5B18, *C. rosea* WL8B28 and *T. harzianum* WL9B49) were associated with antimicrobial activity against *E. coli*. In the EtOAc extract of *A. stygium* WL1B5, two metabolites in particular were proposed: nodulisporone sodium and daidzein. These results confirmed that fungi isolated from polluted environments are promising sources of bioactive secondary metabolites, calling for further exploration as part of the search for new antibiotic drugs.

Author Contributions: Conceptualization, J.V.B.d.S., W.O.P.F.S., É.S.d.S. and A.C.A.C.; methodology, W.O.P.F.S., E.S.L., R.S.d.O., R.M.L., L.A.d.O.; software, W.O.P.F.S., L.A.d.O. and P.A.L.S.; data analysis, W.O.P.F.S. and J.V.B.d.S. investigation, writing—original draft preparation W.O.P.F.S.; writing—review and editing, J.V.B.d.S. and H.F.; visualization, H.F.; supervise, J.V.B.d.S.; project administration, J.V.B.d.S.; funding acquisition, J.V.B.d.S. All authors have read and agreed to the published version of the manuscript.

Funding: We would like to thank the Fundação de Amparo à Pesquisa do Estado do Amazonas (FAPEAM) for the funding of the research by W.O.P.F.S. under the Universal Call FAPEAM-010/2021 CT&I and for the POSGRAD 2020 grant. All the authors also thank the other funding agencies involved: CNPq, CAPES and FINEP. This study was financed in part by the Coordenação de Aperfeiçoamento de Pessoal de Nível Superior—Brasil (CAPES)—Finance Code 001.

Institutional Review Board Statement: Not applicable, because the assessment was solely conducted with microorganisms.

Informed Consent Statement: Not applicable.

Data Availability Statement: All relevant data are presented in the manuscript. Raw data of the study can be provided on reasonable request by the authors.

Conflicts of Interest: The authors declare no conflict of interest.

Appendix A

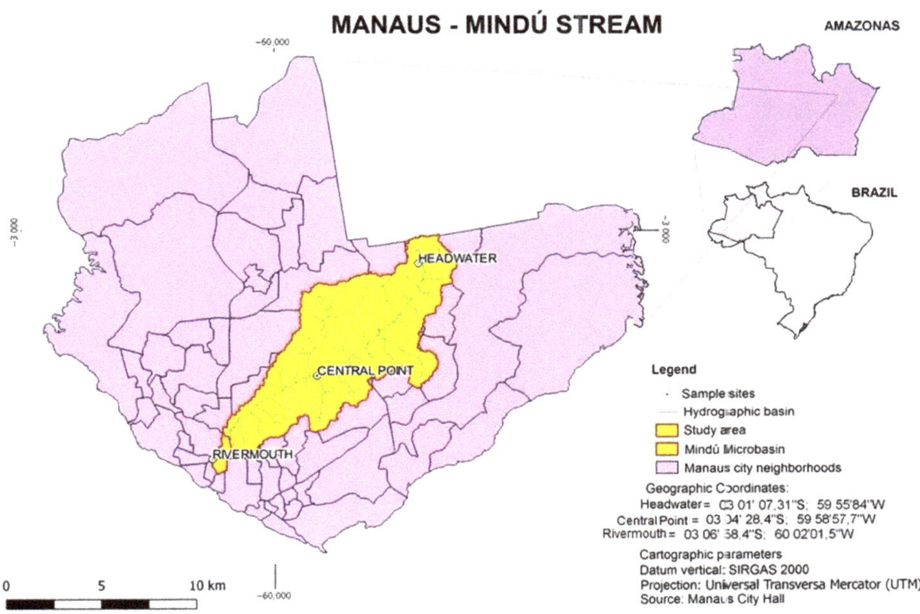

Figure A1. Map of Manaus city (Amazonas, Brazil) showing the research site and sampling points distributed at Mindú stream.

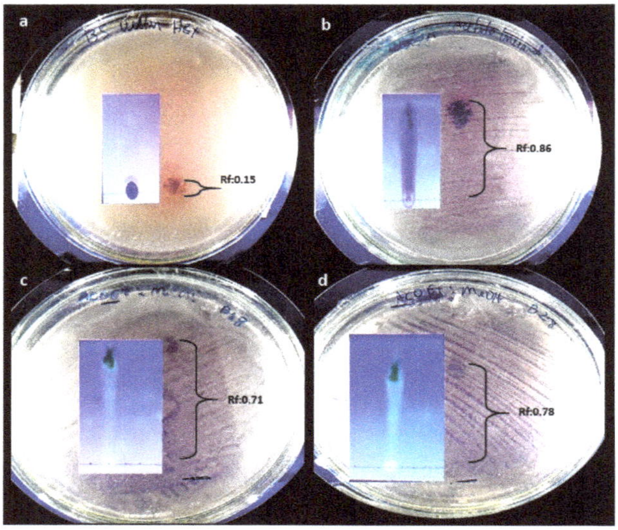

Figure A2. Mobile phase, main bands and the Rf value of the bands that were associated with antibacterial activity against *E. coli* ATCC 25922. *A. stygium* WL1B5 – Hex 100% (**a**), MeOH 100% (**b**); *C. rosea* WL5B18—EtOAc/MeOH 1:1 *v/v* (**c**); *C. rosea* WL8B28—EtOAc/MeOH 1:1 *v/v* (**d**).

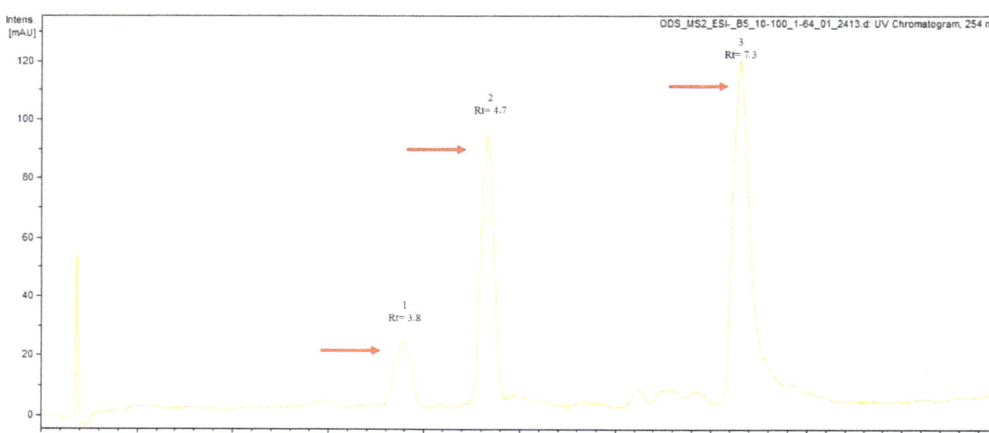

Figure A3. Chromatograms (UV 254 nm) of EtOAc extract of *A. stygium* WL1B5 by LC-MS/MS and the three major peaks at 3.8 (C1), 4.7 (C2) and 7.3 (C3) min. During the research in the literature for information on the peak eluting at 4.7, no equivalent mass compounds were found.

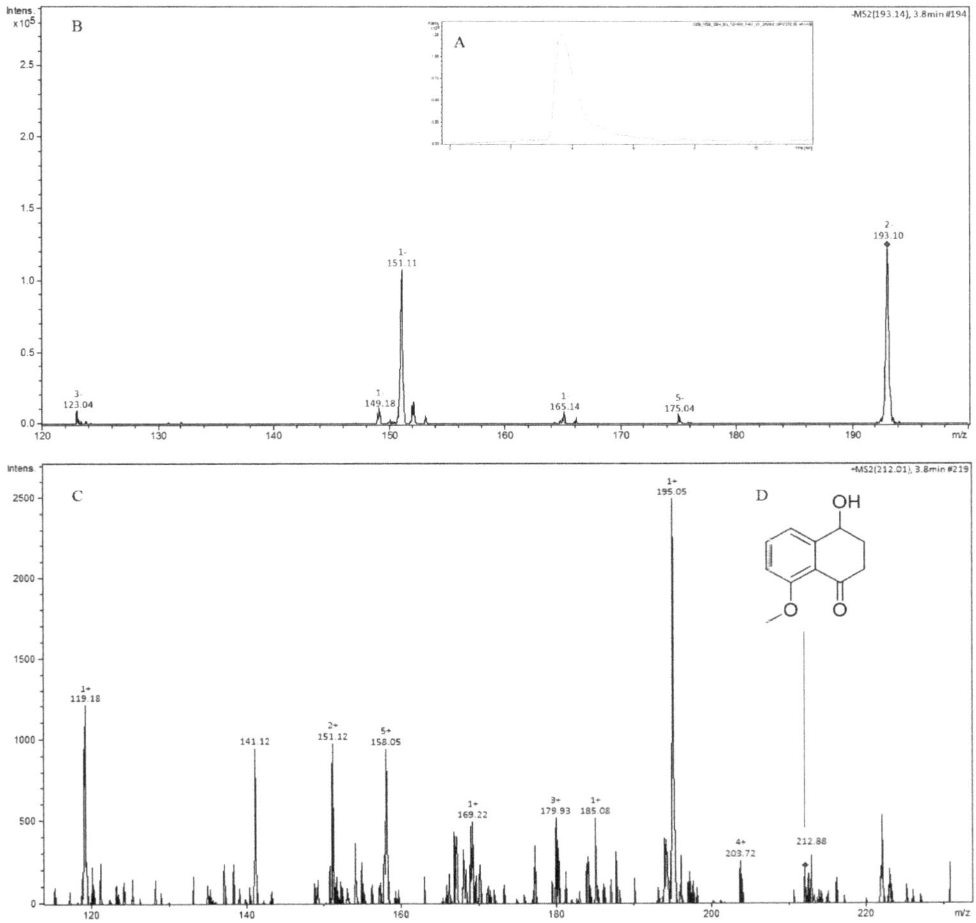

Figure A4. (**A**) Peak of the MS2 spectrum; (**B**) MS2 spectrum of [M − H] *m*/*z* 193 in negative mode; and (**C**) [M + H] *m*/*z* 212 in positive mode of compound 1 from LC-MS/MS analysis of EtOAc extract produced by *A. stygium* WL1B5; (**D**) Nodulisporone. The assessment was performed without replicates.

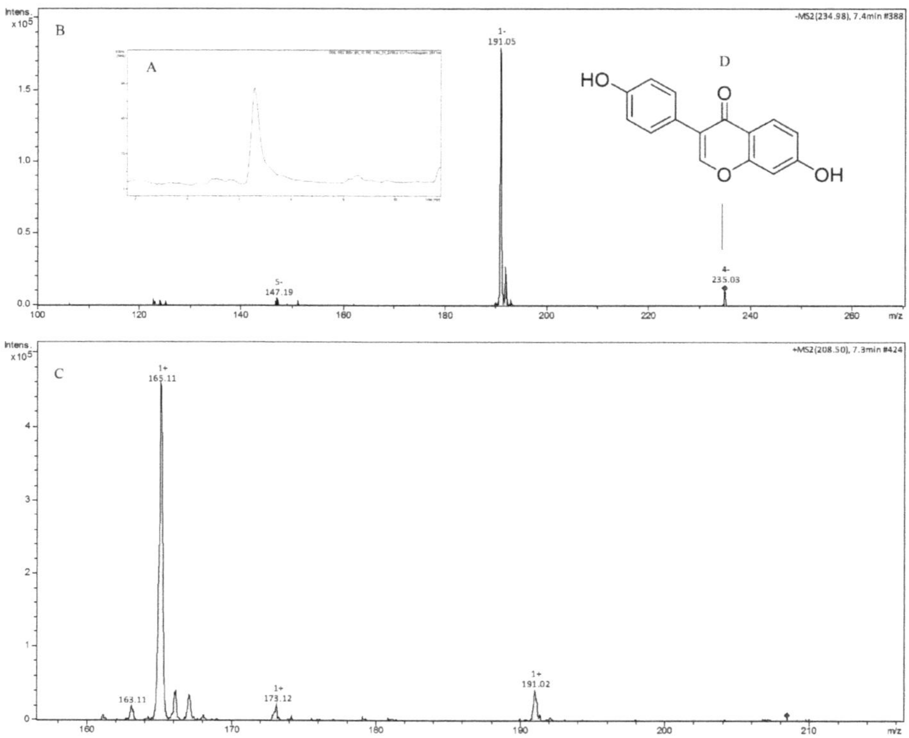

Figure A5. (**A**) Peak of the MS2 spectrum; (**B**) MS2 spectrum of [M − H] m/z 234 in negative mode; (**C**) [M + H] m/z 208 in positive mode of compound 3 from LC-MS/MS analysis of EtOAc extract produced by *A. stygium* WL1B5. (**D**) Daidzein.

References

1. WHO. New Report Calls for Urgent Action to Avert Antimicrobial Resistance Crisis. 2019. Available online: https:/ www.who.int/news/item/29-04-2019-new-report-calls-for-urgent-action-to-avert-antimicrobial-resistance-crisis (accessed c 5 October 2022).
2. Khan, A.; William, R.M.; Cesar, A.A. Mechanisms of antimicrobial resistance among hospital-associated pathogens. *Exper. Re Anti-Infect. Ther.* **2018**, *16*, 269–287. [CrossRef] [PubMed]
3. Nolte, O. Antimicrobial resistance in the 21st century: A multifaceted challenge. *Protein Peptide Lett.* **2014**, *21*, 330–335. [CrossRe [PubMed]
4. Blackwell, M. The fungi: 1, 2, 3 . . . 5.1 million species? *Am. J. Bot.* **2011**, *98*, 426–438. [CrossRef] [PubMed]
5. Smith, H.A. Production of Antimicrobials and Antioxidants from Filamentous Fungi. Doctoral Thesis, National Unive sity of Ireland Maynooth, Kildare, Ireland, 2014. Available online: http://mural.maynoothuniversity.ie/5802/ (accesse on 10 October 2022).
6. Svahn, K.S.; Göransson, U.; El-Seedi, H.; Bohlin, L.; Larsson, D.G.J.; Olsen, B.; Chryssanthou, E. Antimicrobial activity filamentous fungi isolated from highly antibiotic-contaminated river sediment. *Infect. Ecol. Epidemiol.* **2012**, *2*, 11591. [CrossRe
7. Omeike, S.O.; Kareem, S.O.; Lasisi, A.A. Potential antibiotic-producing fungal strains isolated from pharmaceutical waste slud *Beni-Suef Univ. J. Basic Appl. Sci.* **2019**, *8*, 1–7. [CrossRef]
8. Tamura, K.; Nei, M. Estimation of the number of nucleotide substitutions in the control region of mitochondrial DNA in huma and chimpanzees. *Mol. Biol. Evol.* **1993**, *10*, 512–526.
9. Souza Filho, E.A.; Alves, S.B.d.S.M.; Neves, R.K.R.; Batista, I.H.; Albuquerque, C.C.; Damasceno, S.B.; Nascimento, D.A. Estu comparativo de aspectos físico-químicos entre águas da microbacia do mindu e igarapés sob influência antrópica na cidade Manaus-AM. *Braz. J. Dev.* **2020**, *6*, 2419–2433. [CrossRef]
10. Surup, F.; Wiebach, V.; Kuhnert, E.; Stadler, M. Truncaquinones A and B, asterriquinones from Annulohypoxylon truncatu *Tetrahedron Lett.* **2016**, *57*, 2183–2185. [CrossRef]

11. Honório, A.E. Avaliação química e biológica de fungos endofíticos associados às algas marinhas Acanthophora spicifera, Dichotomaria marginata e Sargassum vulgare. Doctoral Thesis, Universidade Estadual Paulista, São Paulo, Brazil, 2018. Available online: https://repositorio.unesp.br/handle/11449/180427 (accessed on 10 October 2022).
12. Medina, R.P.; Araujo, A.R.; Andersen, R.J.; Soares, M.A.; Silva, F.D.A.; Silva, D.H. Aromatic compounds produced by endophytic fungi isolated from red alga Asparagopsis taxiformis - Falkenbergia stage. *Nat. Prod. Res.* **2018**, *3*, 443–446. [CrossRef]
13. Moura, S.S.T. Avaliação da atividade antimicrobiana de fungos endofíticos isolados de plantas no Brasil contra patógenos humanos. Doctoral Thesis, Universidade Federal de Santa Catarina, Florianopolis, Brazil, 2020.
14. Zhai, M.M.; Li, J.; Jiang, C.X.; Hou, Y.; Shi, Y.P.; Di, D.L.; Zhang, J.W.; Wu, Q.X. Isolation of Secondary Metabolites from the Soil-Derived Fungus Clonostachys rosea YRS-06, a Biological Control Agent, and Evaluation of Antibacterial Activity. *J. Agric. Food Chem.* **2016**, *61*, 2298–2306. [CrossRef]
15. Dos Santos Dias, A.C.; Couzinet-Mossion, A.; Ruiz, N.; Lakhdar, F.; Etahiri, S.; Bertrand, S.; Ory, L.; Roussakis, C.; Pouchus, Y.F.; Nazih, E.H.; et al. Steroids from marine-derived fungi: Evaluation of antiproliferative and antimicrobial activities of eburicol. *Mar. Drugs* **2019**, *17*, 372. [CrossRef]
16. Han, P.; Zhang, X.; Xu, D.; Zhang, B.; Lai, D.; Zhou, L. Metabolites from Clonostachys fungi and their biological activities. *J. Fungi* **2020**, *6*, 229. [CrossRef] [PubMed]
17. Cheng, M.J. Secondary metabolites from the endophytic fungos Annulohypoxylon stygium BCRC 34024. *Chem. Nat. Compd.* **2014**, *50*, 237–241. [CrossRef]
18. Gao, S.; Tian, W.J.; Liao, Z.J.; Wang, G.H.; Zeng, D.Q.; Liu, X.Z.; Wang, X.Y.; Zhou, H.; Chen, H.F.; Lin, T.; et al. Chemical Constituents from Endophytic Fungus Annulohypoxylon cf. stygium in Leaves of Anoectochilus roxburghii. *Chem. Biodivers.* **2020**, *17*, 200424. [CrossRef] [PubMed]
19. Abdel-Wahab, N.M.; Harwoko, H.; Müller, W.E.; Hamacher, A.; Kassack, M.U.; Fouad, M.A.; Proksch, P. Cyclic heptapeptides from the soil-derived fungus Clonostachys rosea. *Bioorg. Med. Chem.* **2019**, *27*, 3954–3959. [CrossRef]
20. Sofian, F.F.; Suzuki, T.; Supratman, U.; Harneti, D.; Maharani, R.; Salam, S.; Abdullah, F.F.; Koseki, T.; Tanaka, K.; Kimura, K.; et al. Cochlioquinone derivatives produced by coculture of endophytes, Clonostachys rosea and Nectria pseudotrichia. *Fitoterapia* **2021**, *155*, 105056. [CrossRef]
21. Kamisuki, S.; Ishimaru, C.; Onoda, K.; Kuriyama, I.; Ida, N.; Sugawara, F.; Yoshida, H.; Mizushina, Y. Nodulisporol and Nodulisporone, novel specific inhibitors of human DNA polymerase λ from a fungus, *Nodulisporium* sp. *Bioorg. Med. Chem.* **2007**, *15*, 3109–3114. [CrossRef]
22. Hayasaka, S.; Koseki, T.; Murayama, T.; Kwon, E.; Shiono, Y. Phenylisobenzofuranones from Fungicolous Nodulisporium sp. SH-1. *Zeitschrift für Naturforsch B* **2011**, *66*, 961–964. [CrossRef]
23. Wang, Q.; Wang, H.; Xie, M. Antibacterial mechanism of soybean isoflavone on staphylococcus aureus. *Arch. Microbiol.* **2010**, *192*, 893–898. [CrossRef]
24. Singhvi, M.S.; Zinjarde, S.S. Production of pharmaceutically important genistein and daidzein from soybean flour extract by using β-glucosidase derived from Penicillium janthinellum NCIM 1171. *Process. Biochem.* **2020**, *97*, 183–190. [CrossRef]
25. Dang, H.; Kong, Q.; Winchester, W.; Wan, X.; Lei, Y.; Zhang, H.; Zhao, Y.; Liu, X.; Xu, B.B.; Zhang, B.; et al. Isolation, Identification and Pathogenic Effects of Trichoderma spp. from *Auricularia auricula. Res. Sq.* 2022; *(preprint)*.
26. Xuan, Q.C.; Huang, R.; Miao, C.P.; Chen, Y.W.; Zhai, Y.Z.; Song, F.; Wang, T.; Wu, S. Secondary Metabolites of Endophytic Fungus Trichoderma Sp. YM 311505 Of Azadirachta Indica. *Chem. Nat. Compd.* **2014**, *50*, 139–141. [CrossRef]
27. Guo, L.; Wang, C.; Zhu, W.C.; Xu, F.Q. Bioassay-guided fractionation and identification of active substances from the fungus Aspergillus tubingensis against Vibrio anguillarum. *Biotechnol. Biotechnol. Equip.* **2016**, *30*, 602–606. [CrossRef]
28. Xie, Y.; Yang, W.; Tang, F.; Chen, X.; Ren, L. Antibacterial activities of flavonoids: Structure-activity relationship and mechanism. *Curr. Med. Chem.* **2015**, *22*, 132–149. [CrossRef] [PubMed]
29. APHA. *Standard Methods for the Examination of Water and Wastewater*; APHA: Washingt, DC, USA, 2005.
30. Ferrer, C.; Coloms, F.; Frasés, S.; Mullet, E.; Abad, J.L.; Alió, J.L. Detection and identification of fungal pathogens by PCR and by ITS2 and 5.8S ribosomal DNA typing in ocular infections. *J. Clin. Microbiol.* **2001**, *39*, 2873–2879. [CrossRef]
31. White, T.J.; Bruns, T.; Lee, S.; Taylor, J. PCR Protocols: A Guide to Methods and Applications. In *Amplification and Direct Sequencing of Fungal Ribosomal RNA Genes for Phylogenetics*; Innis, N., Gelfand, D., Sninsky, J., White, T., Eds.; Academic Press: New York, NY, USA, 1990; pp. 315–322.
32. O'donnell, K. The Fungal Holomorph; Mitotic, Meiotic and Pleomorphic Speciation in Fungal Systematic. In *Fusarium and Its Near Relatives*; CAB International: Wallingford, UK, 1993; pp. 225–233.
33. Kumar, S.; Stecher, G.; Li, M.; Knyaz, C.; Tamura, K. MEGA X: Molecular evolutionary genetics analysis across computing platforms. *Mol. Biol. Evol.* **2018**, *35*, 1547–1549. [CrossRef]
34. Lima, A.M.D.; Salem, J.I.; Souza, J.V.B.D.; Cortez, A.C.A.; Carvalho, C.M.; Chaves, F.C.M.; Junior, V.F.d. Effects of culture filtrates of endophytic fungi obtained from Piper aduncum L. on the growth of mycobacterium tuberculosis. *Electron. J. Biotechnol.* **2011**, *14*, 8.
35. *CLSI Document M02-A11*; Performance Standards for Antimicrobial Disk Susceptibility Tests; Approved Standard—Eleventh Edition. Clinical and Laboratory Standards Institute: Wayne, PA, USA, 2012.
36. *CLSI document M07-A9*; Performance Standards for Antimicrobial Disk Susceptibility Tests; Approved Standard—Eleventh Edition. Clinical and Laboratory Standards Institute: Wayne, PA, USA, 2012.

37. Ahmed, S.A.; Gogal, J.R.; Walsh, J.E.A. A new rapid and simple non- radioactive assay to monitor and determine the proliferation of lymphocytes an alternative to [3H]thymidine incorporation assay. *J. Immunol. Methods.* **1994**, *170*, 211–224. [CrossRef]
38. Khurram, M.; Khan, M.A.; Hameed, A.; Abbas, N.; Qayum, A.; Inayat, H. Antibacterial activities of Dodonaea viscosa using contact bioautography technique. *Molecules* **2009**, *14*, 1332–1341. [CrossRef]

antibiotics

MDPI

eview

aboratory Surveillance, Quality Management, and Its Role in ddressing Antimicrobial Resistance in Africa: Narrative Review

halid Musa [1,2,*], Ijeoma Okoliegbe [1], Tassabeeh Abdalaziz [1,2], Ahmed Taha Aboushady [3], John Stelling [3] nd Ian M. Gould [1]

[1] Department of Medical Microbiology and Virology, Aberdeen Royal Infirmary, Aberdeen AB25 2ZN, UK; ijeoma.okoliegbe@nhs.scot (I.O.); tassabeeh.abdalaziz@nhs.scot (T.A.); ian.gould@nhs.scot (I.M.G.)

[2] Department of Acute Medicine and Infectious Diseases, Aberdeen Royal Infirmary, Aberdeen AB25 2ZN, UK

[3] Division of Infectious Diseases, Brigham and Women's Hospital, Harvard Medical School Boston, MA 02115, USA; aaboushady@whonet.org (A.T.A.); jstelling@whonet.org (J.S.)

* Correspondence: khalid.musa@nhs.scot

Abstract: AMR is a major public health concern that calls for extensive work and a multidisciplinary team approach. The high prevalence of infectious diseases in African nations leads to widespread antibiotic usage and eventual antimicrobial resistance, which has significant negative effects on people's health, the economy, and society. Additionally, inadequate or nonexistent antimicrobial drug regulations, inappropriate prescription practices, and restrictions on public health prevention initiatives such as immunization, water and sanitation, and sexual health may all contribute to the emergence of AMR. Despite the need for laboratory quality assurance, many African laboratories confront substantial difficulties in implementing efficient quality assurance programs. AMR surveillance in Africa is hampered by a lack of laboratory capacity, insufficient data collection and analysis, and poor stakeholder collaboration. Several initiatives and programs, including the World Health Organization's Global Antimicrobial Resistance and Use Surveillance System (GLASS), the Africa Centres for Disease Control and Prevention (Africa CDC) Antimicrobial Resistance Surveillance Network (AMRSNET), and the Fleming Fund, a UK government initiative aimed at tackling AMR in low- and middle-income countries, have been established to strengthen AMR surveillance in Africa and globally.

Keywords: antimicrobial resistance; AMR; antimicrobial resistance in Africa; quality management; AMR surveillance

1. Introduction

Antibiotics serve a key role in lowering the global burden of communicable illnesses. However, the ability of pharmaceuticals to cure infectious diseases and the efficacy of drugs are not limitless. AMR jeopardizes the efficacy of infection treatment and is a public health issue with national and global implications. The spreading AMR pandemic is a global public health problem with far-reaching health, economic, and societal consequences [1]. Resistance develops as a result of selection pressure from rational and indiscriminate antimicrobial usage in human health, veterinary medicine, agriculture, and the environment. Infections produced by resistant bacteria result in prolonged illness duration, greater fatality rates, and increased expenses associated with alternative treatment, and AMR is seen as a global economic threat [2].

One of the biggest concerns to public health in the twenty-first century is AMR. It is critical to have knowledge about the top pathogen-drug combinations contributing to the burden of bacterial AMR, trends around the world, and the current size of the burden. If AMR continues to grow, many infections could become far more deadly than they are today [3].

Among African nations, communicable diseases continue to be the largest cause of death, accounting for the great majority of fatalities among children under the age of five. In comparison to other World Health Organization (WHO) areas, Africa has a much greater level of poverty and communicable disease risk factors, which are highly correlated with the number of infectious illness cases recorded [4].

AMR is a significant worldwide health problem that necessitates extensive work and a diverse team approach. Tracking AMR trends, patterns, treatments, and policy responses can be performed with the help of surveillance. The formulation of logical actions and policies is made easier by the proper correlation of surveillance data, which provide insightful information about the underlying epidemiology [5]. This paper's goal is to increase the awareness of AMR surveillance in Africa and to highlight the existing situation of AMR in Africa.

2. Antimicrobial Resistance in Africa

The high prevalence of infectious diseases in African nations leads to widespread antibiotic usage and eventual antimicrobial resistance, which has significant negative effects on people's health, the economy, and society. The lack of availability or access to antimicrobial treatments, as well as the possibility of low-quality or counterfeit medications when they are accessible, are the key factors aggravating the AMR problem in Africa. Additionally, antimicrobials, including antibiotics, can be purchased over the counter or fed to animals as preventative medicine or growth promoters in low-resource environments such as Africa [4].

Despite the fact that antimicrobial resistance emerges and spreads naturally, there is a higher selection pressure that is primarily brought on by the overuse and abuse of antibiotics in both human and animal medicine. In developing nations where strict rules on antibiotic use are not entirely in place and implemented, livestock production is rising. Large quantities of antibiotics, such as penicillin, β-lactams, fluoroquinolones, and aminoglycosides, are used by farmers on their livestock. Additionally, sub-therapeutic levels are used in animal feeds to promote productivity, reduce illness, and promote growth. High rates of resistance to the main antimicrobials used in human treatments in bacteria isolated from food-producing animals have been linked to the abuse of antibiotics. Bacteria isolated from plants and the environment have high reported rates of resistance [6].

The majority of low-resource settings still deal with outdated facilities, a lack of technical expertise, and a shortage of essential materials for AMR diagnosis and treatment. In low- and middle-income (LMIC) nations, this results in an increase in infectious diseases and AMR. Even in some national reference laboratories, AMR detection in LMICs is poor, which is indicative of a lack of laboratory capacity to support diagnostic processes [7].

The WHO announced a priority list for discovering novel and effective antibiotic therapies in 2017 [8]. The list was created to help guide research and development priorities for new antibiotics, with a focus on bacteria with multidrug resistance that cause serious and frequently fatal infections in healthcare and nursing home settings. Although the goal of this list was to establish new antibiotic studies and development priorities instead of identifying the most challenging pathogen–drug combinations, its utility in determining priorities has been limited by the lack of an ongoing nationwide evaluation concerning the burden of bacterial AMR [8].

In 2018, the African Union (AU) formed a Task Force on Antimicrobial Resistance in conformity with the United Nations (UN) strategy and to fulfil the specific needs of African nations. This Task Force represents AU agencies working in the human, animal, and plant health sectors to measure, prevent, and minimize the effects caused by AMR bacteria [9].

AMR typically occurs in microorganisms that cause pneumonia, diarrheal illnesses, tuberculosis (TB), sexually transmitted diseases, and malaria in Africa. These pathogens are among those that are likely to be spread within the population. Fighting TB, malaria, pediatric dysentery, and pneumonia now costs much more money due to drug resistance. Additionally, it jeopardizes efforts to successfully treat HIV/AIDS patients [10]. In a recent attempt to define the AMR burden in Africa, the WHO African region carried out a systematic review comprising isolates from 2016 to 2020. They concluded that *Escherichia coli* (*E. coli*) was the most resistant Gram-negative bacteria with more than 20% resistance observed for amoxicillin (24.5%), ampicillin (23.5%), and trimethoprim/sulfamethoxazole (22.5%). Additionally, high resistance was observed for other first- and second-line antibiotics; for example, amoxicillin/clavulanic acid (13.2%), chloramphenicol (12.3%), and ciprofloxacin (8.2%), (Figure 1) [7]. Similarly for Gram-positive bacteria, of worrying concern is the high rate of resistance of *Streptococcus pneumoniae* to trimethoprim/sulfamethoxazole (64.3%) which is a commonly used antibiotic. Other commonly used antibiotics with a high percentage of resistance include oxacillin (32.2%), tetracycline (28.3%), penicillin (23.2%), amoxicillin (20.6%), and ampicillin (19.3%) (Figure 1) [7].

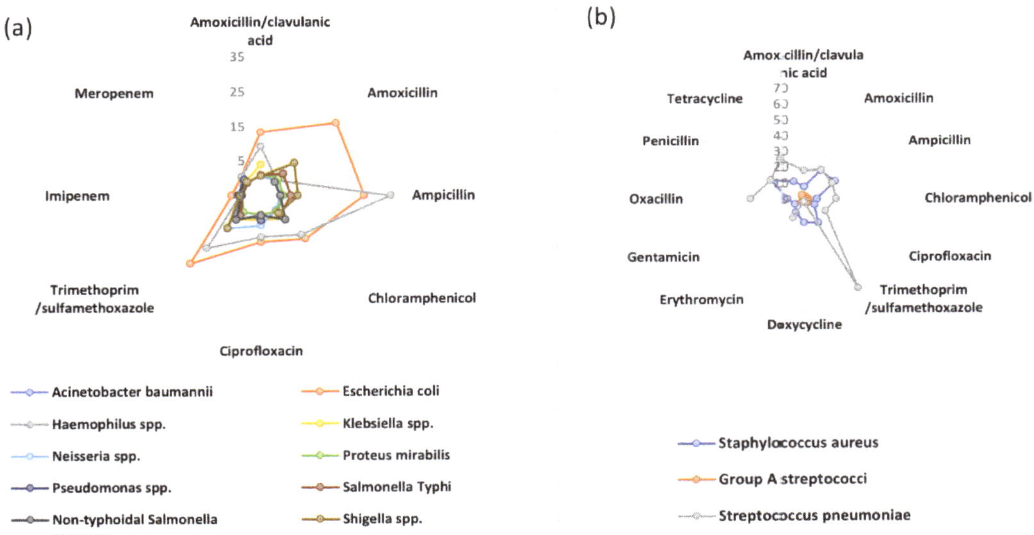

Figure 1. AMR patterns among (**a**) Gram−negative and (**b**) Gram−positive bacteria in the WHO African Region, 2016–2020 to first- and second-line antibiotics.

A proper laboratory-based surveillance program is necessary for identifying resistance and tracking its dissemination. Therefore, in order to extend the shelf life of antimicrobial agents in African countries, it is necessary to improve a number of factors, including access to diagnostic laboratories, surveillance of the emergence of resistance, regulation, and public, clinician/prescriber, and veterinary education regarding the proper use of antibiotics [4]. Implementing efficient and long-lasting AMR surveillance initiatives is difficult in low- and middle-income nations, such as those in Africa, in contrast to high-income nations. There is aa lack of institutional and infrastructure capabilities, a lack of funding and human resources, the underutilization of data already accessible, and limited transmission to regulatory agencies [7].

Poor AMR surveillance data over the entire continent of Africa make it difficult to comprehend the true scope and impact of AMR. When available, country data are not consistently gathered, shared with, or acknowledged by national entities, which restricts their power to influence national activities [11]. In response to this neglect, the 68th World Health Assembly (WHA) established a Global Action Plan (GAP) to address AMR with the overriding objective of raising awareness of AMR on a national and international level. The first two GAP-proposed goals, which center on raising public knowledge and comprehension of AMR through study and surveillance, are among a group of goals. Despite the GAP policy's recommendation for the creation of national action plans (NAPs) and ongoing surveillance of priority pathogens, only two countries in the African region had adopted NAPs for AMR, and none had implemented any kind of national surveillance [6].

Antimicrobial resistance, which is presently thought to be responsible for more than 700,000 deaths annually worldwide, is causing increasing worry on a global scale. By 2050, if suitable steps are not taken to slow its spread, AMR will be responsible for nearly 10 million deaths and USD 100 trillion in annual costs [12]. Routine surveillance is therefore essential, especially in LMICs and Africa where AMR is expected to have the greatest impact [6]. As opposed to other health issues, AMR impacts every nation, regardless of its wealth or level of development, because resistant microbes know no national boundaries.

3. Quality Management

Effective disease prevention and management relies in accurate and time-efficient laboratory testing for patient care [13]. Healthcare systems are tasked with continuous quality improvement of patient care and safety [14]. High-quality microbiological diagnostics are essential for clinicians to make accurate diagnoses, formulate treatment plans, and subsequently monitor treatment effects on their patients [13–17]. Therefore, in developed countries laboratory diagnostics play a crucial role at every stage of disease management with clinicians relying on it for the diagnosis, monitoring, screening, and prognosis of disease. Here, information gathered through diagnostics, surveillance, monitoring, or follow-ups contributes to an understanding of the clinical picture, highlights opportunities for optimal patient care and improves the quality of care during the patient's journey. This efficient use of laboratory test results in patient management results in the prudent use of antibiotics while reducing unnecessary treatments, improving diagnosis accuracy, monitoring disease progression, and evaluating treatment efficacy [18]. Miller et al. [19] proposed that the best outcomes for patients occur because of a strong partnership between the clinician and the laboratory. However, in Africa with scarce resources and skills required to perform accurate and reliable microbiology, most laboratories are often under-resourced, if they exist at all. Thus, with the nearly nonexistent quality assurance and control procedures, clinicians then resort to empirical antibiotics treatments or are reluctant to use laboratory tests in the management of patients, especially when results are discordant with clinical diagnosis [20]. In a cross-sectional study of 283 participants conducted in 2018 in Nigeria which assessed the reasons for poor utilization of laboratory testing in definitive diagnosis and infection management, several factors such as the physicians' perception and attitude of the relevance of the laboratory as well as the apparent inadequacy of microbiology laboratory practices were proposed [21]. In a similar study, other challenges that affected the utilization of laboratory test results include delayed test results, inadequate test menus, a lack of feedback mechanisms, and poor quality assurance systems. In the author's experience, even such basics as an adequate supply of the correct antibiotic discs for susceptibility can be a problem. Aid programs may supply antibiotics to clinicians but not think of the laboratory testing issues. Similarly, automated systems may be supplied in aid programs but there is no capacity for maintenance and repairs during break downs.

Laboratory quality assurance refers to the processes and procedures that laboratories use to ensure the accuracy and reliability of their test results. In Africa, laboratory quality assurance is an important tool for improving the quality of laboratory services and ensuring the reliability of test results. Despite the importance of laboratory quality assurance, many laboratories in Africa face significant challenges in implementing effective quality assurance programs. These challenges include a lack of resources, inadequate infrastructure, and limited access to training and technical support. To address these challenges, several organizations are working to support laboratories in Africa in their efforts to improve their quality assurance practices. One such organization is the African Society for Laboratory Medicine (ASLM), which promotes the use of the Stepwise Laboratory Quality Improvement Process Towards Accreditation (SLIPTA) framework [22]. SLIPTA provides a structured approach for laboratories to assess their current level of quality and identify areas for improvement using a star criteria (Figure 2). The framework includes a focus on quality assurance practices such as internal and external quality control, proficiency testing, and continuous quality improvement [22]. Countries are graded using the star criteria.

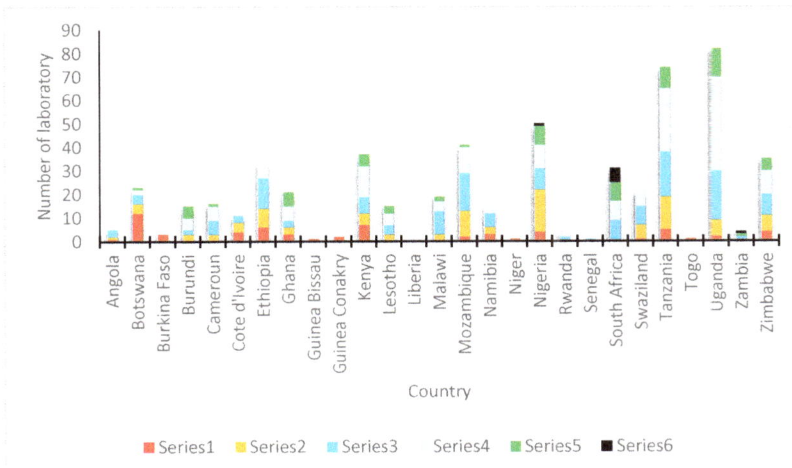

Figure 2. Star criteria used by SLIPTA to assess current level of quality and to identify areas for improvement for laboratories in Africa.

In addition to ASLM, other organizations such as the World Health Organization Regional Office for Africa (WHO/AFRO) and the US Centers for Disease Control and Prevention (CDC) are also working to support laboratory quality assurance in Africa. These organizations provide training, technical assistance, and other resources to help laboratories improve their quality assurance practices and achieve accreditation. Strengthening Laboratory Management Toward Accreditation (SLMTA) was developed as a task-based training and mentoring toolkit that provides laboratory personnel with the skills and knowledge they need to implement quality management systems [23]. SLMTA uses a multi-workshop implementation model to deliver training to laboratory managers. The training is composed of several tasks designed as tools and job aids which aid laboratory managers to enhance their management routines. Additionally, the program empowers managers with tools to initiate laboratory improvement measures with existing resources leading to better patient care and public health outcomes [23]. These programs have been successful in helping many laboratories improve their quality management systems with some African laboratories achieving the International Organization for Standardization (ISO) 15,189 accreditation (Figure 3).

Despite its successes, there are still challenges associated with its use in Africa. One challenge is the inherent lack of resources and infrastructure which limits the capacity and performance of laboratories and hinders their ability to comply with SLIPTA requirements. Additionally, poor communication, coordination, and integration among different levels of laboratory services can hinder the progress of these programs while insufficient political commitment by the local stakeholders, such as ministries of health, financial support, monitoring and evaluation, feedback mechanisms, and incentives for quality improvement result in a lack of sustainability. In addition to these challenges, there may also be cultural and organizational barriers to implementing SLIPTA. For example, some laboratory staff may be resistant to change or may not fully understand the importance of quality management systems. In addition, some laboratories may lack the leadership and support needed to successfully implement these programs. Finally, there is an urgent need for harmonization and alignment of SLIPTA with other existing or emerging laboratory quality initiatives in Africa, such as ASLM, SLMTA, and the International ISO 15189. This would help to avoid confusion, duplication, and contradiction among different standards, guidelines, and tools for laboratory quality improvement [24].

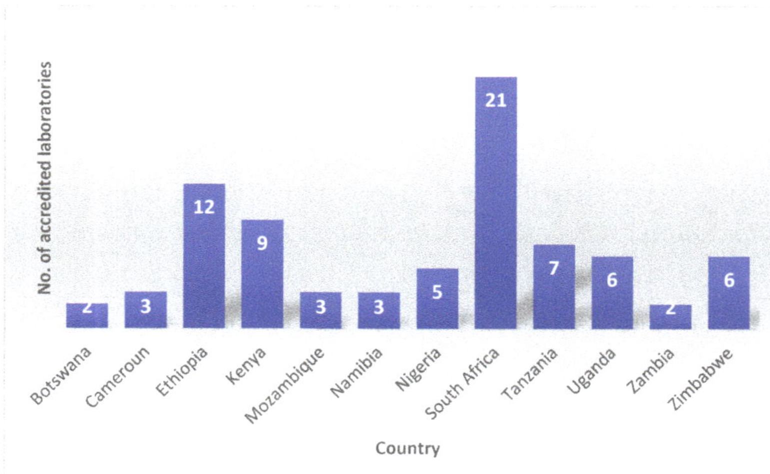

Figure 3. African laboratories achieving the International Organization for Standardization (ISO) 15,189 accreditation.

4. AMR Surveillance

Antimicrobial resistance surveillance has become an essential component of global health strategies to address AMR's significant global threat. Surveillance plays a crucial role in guiding policies and infection prevention and control measures. Its significance lies in assessing the spread of AMR and monitoring the effectiveness of local, national, and global strategies. AMR surveillance involves monitoring trends and resistance patterns in different geographic regions, populations, and bacterial strains. This information is critical for identifying emerging resistance patterns, informing clinical practice and treatment guidelines, and guiding the development and deployment of new antimicrobial agents.

Several initiatives and programs have been established to strengthen AMR surveillance in Africa and globally, including the World Health Organization's Global Antimicrobial Resistance and Use Surveillance System (GLASS) [25], the Africa Centres for Disease Control and Prevention (Africa CDC) Antimicrobial Resistance Surveillance Network (AMRSNET) [26], and the Fleming Fund, a UK government initiative aimed at tackling AMR in low- and middle-income countries [27]. These programs strengthen the national capacity to collect data on the prevalence and patterns of AMR from healthcare facilities, laboratories, and community settings, providing valuable insights into the current status of AMR and identifying areas where interventions are needed.

AMR surveillance in Africa faces several challenges, including limited laboratory capacity, inadequate data collection and analysis, and poor stakeholder coordination. However, with the growing recognition of AMR's threat, there is increasing momentum to strengthen surveillance systems and improve data sharing and collaboration across the continent [28]. According to the Tripartite AMR Country Self-Assessment Survey (TrACSS) in 2022, only 16 of the 42 responding countries in the WHO African region use the relevant antimicrobial resistance surveillance data to inform operational decision-making and amend policies [29]. Furthermore, 70% have National Action Plans for AMR that are being implemented, and over half of them have already costed and budgeted their operational plans and adopted a monitoring mechanism [29].

One of the major obstacles is the insufficient demand for bacteriology diagnostics among healthcare providers because of their high costs and turnaround time. Moreover, health systems do not prioritize the supplies necessary for bacteriology diagnostics, which results in a consistent shortage of human resources and interruptions in the supply chain. Additionally, laboratory personnel may hesitate to conduct bacteriology diagnostics due to their time-consuming and inconvenient nature [30].

To ensure effective laboratory-based surveillance, it is essential to have a solid infectious disease diagnostic cycle, which involves clinicians sending samples for culture and sensitivity testing, a laboratory that can generate quality reports, a reporting system that delivers timely results to the clinician who requested the test, and a functional laboratory information system (LIS) that can provide information to the surveillance program. To have any functional surveillance system in the human health sector, it is necessary to establish at least one national reference laboratory in each country that can conduct phenotypic and genotypic evaluations of resistance and resistance genes [30]. Only Cabo Verde and São Tomé and Principe do not have a reference laboratory performing antimicrobial susceptibility testing [29].

4.1. Types of AMR Surveillance

For AMR surveillance, there are two approaches: active and passive surveillance. These methods are utilized to monitor and track the prevalence of AMR. Each method presents distinct advantages and disadvantages, and as such, a combination of both can offer a more comprehensive understanding of the dynamics of AMR [31].

Active surveillance is a method that involves proactively seeking out cases of AMR within a population. This can be achieved by systematically sampling patients or conducting focused surveillance among high-risk groups, such as hospital patients or individuals working with livestock. Active surveillance can provide a more nuanced comprehension of the prevalence of AMR and the emergence of resistant strains. A notable benefit of this approach is that it enables the collection of reliable data that can inform public health policies and guide the development of targeted interventions. Nonetheless, active surveillance can be resource-intensive and challenging to sustain over an extended period, particularly in resource-limited settings [32].

On the other hand, passive surveillance involves collecting data on AMR cases identified through routine clinical testing or other diagnostic procedures. This method relies on healthcare providers and laboratories to report cases of AMR to public health authorities. Passive surveillance can provide a more comprehensive view of a population's AMR trends over time and help identify changes in resistance patterns. In contrast to active surveillance, passive surveillance is less resource-intensive and can be integrated into routine clinical practice. However, passive surveillance may be subject to underreporting, mainly if healthcare providers are not aware of the importance of reporting AMR cases or if reporting is not mandatory [33].

Active and passive surveillance have limitations, but when used together, they can provide a complete picture of the prevalence and dynamics of AMR in a population. This information can guide public health interventions, inform treatment decisions, and guide the development of new antibiotics and other therapies. It is essential to distinguish between the terms "active" and "passive" in antimicrobial resistance (AMR) surveillance, as they are often used inaccurately and inconsistently. For instance, performing whole-genome sequencing on routine isolates is considered passive surveillance, despite active testing being involved. Genuine active surveillance programs are rare, whereas there is a wide range of passive surveillance programs, including routine data collection or gathering additional patient information, conducting further tests on isolates, and implementing new alerts and actions. Most national AMR surveillance programs describe themselves as "active" even though they still rely on routine clinical samples, rendering them not entirely active according to traditional definitions.

In the WHO African region, among the 42 countries that participated in the TrACSS survey in 2022 [29], 14 countries possess a standardized national system for monitoring antimicrobial resistance (AMR). These countries collect data on common bacterial infections in both hospitalized and community patients. They have also established a network of surveillance sites, a designated national reference laboratory for AMR, and a national coordinating center responsible for generating reports on AMR. On the other hand, 20 countries compile their data at the local level, while 4 countries gather their data nationally. However, these countries lack a standardized approach, national coordination, and/or quality management. Nonetheless, Lesotho, Central African Republic, and South Sudan face challenges in generating data on AMR.

4.2. Genomic Surveillance

Genomic surveillance of AMR entails the use of state-of-the-art sequencing technologies to investigate the genomic evolution and dissemination of antibiotic-resistant bacteria. This approach involves the meticulous analysis of bacterial strains' genetic material to pinpoint specific resistance mechanisms and mutations responsible for AMR. Given its capacity to promptly identify newly emerging resistant strains and facilitate the development of more precise treatment and preventive measures, this approach is gaining prominence in the fight against AMR.

Genomic surveillance of AMR utilizes multiple techniques, with whole-genome sequencing (WGS) being a prevalent method. WGS entails sequencing the complete genome of a bacterial strain to pinpoint genetic mutations linked to AMR. This technique yields comprehensive data on the genetic mechanisms underlying resistance and enables the detection of new or developing strains of resistant bacteria [34]. Another approach is metagenomics, where all genetic material in a given sample, including bacterial DNA, is sequenced. This technique can effectively detect resistant bacteria in intricate settings such as hospitals or farms and furnish valuable information on the dissemination of resistant strains within these environments [35].

Genomic surveillance has several benefits over traditional surveillance methods, such as culturing bacteria on agar plates, which can be time-consuming and may miss essential strains. Genomic surveillance is also more sensitive and specific, allowing for detecting resistance mechanisms that may not be visible with traditional methods [36].

Genomic surveillance is gaining traction in Africa as a tool for preventing and controlling emerging infections, as seen in the 2018 outbreak of Lassa fever virus in Nigeria and Ebola virus outbreaks in the Democratic Republic of the Congo (2018–2020) [37,38]. Genomic data were used with other data to determine transmission dynamics, assess spatiotemporal aspects of the epidemic, and evaluate the efficacy of diagnostic tests and vaccines [39]. Thus, the capacities exist on the content and can be repurposed for antimicrobial resistance.

The data generated through genomic surveillance can inform public health policy and guide the development of new antibiotics and other treatments for resistant infections. For example, the data can help identify high-risk populations, such as patients in hospitals or individuals who work with livestock, and inform targeted intervention strategies [39].

4.3. The Need for a One-Health Surveillance System

One Health is gaining attention in Africa for AMR surveillance due to the high burden of infectious diseases, poor health infrastructure, and weak regulatory systems. One Health aims to ensure the wellbeing of humans, animals, and the environment. It is a comprehensive strategy that emphasizes prevention, prediction, detection, and response to prevent global health crises such as the COVID-19 pandemic. This approach requires collaboration across various sectors and communities to address the underlying causes of health issues and develop sustainable solutions. One Health is relevant to public health, veterinary medicine, environmental protection, and other sectors. It is particularly applicable to ensuring food and water safety, promoting proper nutrition, controlling animal-to-human diseases, managing pollution, and addressing antibiotic-resistant microbes. According to TrACSS 2022, only 19 countries have established or started implementing an integrated multi-sectoral surveillance system for antimicrobial resistance [29].

The African Union Inter-African Bureau for Animal Resources (AU-IBAR) has established the One Health Coordination Group on Zoonotic Diseases program and developed a framework for One Health practice based on collaboration between various stakeholders. The framework emphasizes the importance of collecting and sharing data on AMR in humans, animals, and the environment. A One Health approach to AMR surveillance in Africa is critical for addressing the increasing threat of drug-resistant infections in both humans and animals. Additionally, to support these integration efforts, SILAB for Africa (SILABFA), a surveillance system widely used for laboratory diagnostic activities in animal sectors, is now working on the integration of the human data, specifically through the integration and enhancing the interoperability with WHONET, a software widely used for AMR surveillance in human laboratories [40].

Effective implementation of One Health surveillance strategies for antimicrobial resistance (AMR) in Africa is hindered by limited inter-sectoral collaboration, communication, and coordination among the animal health, environmental health, and human health sectors. Silos and insufficient data sharing due to issues of ownership and confidentiality also pose a significant challenge to One Health surveillance.

Currently, after excluding the four countries with no NAPs, 32 out of the 38 countries developing or implementing their NAPs are ensuring an integrated AMR national planning process with other existing action plans and/or strategies, according to TrACSS 2022, and 23 countries are already implementing AMR awareness-raising activities with other sectors and stakeholders [29].

5. Future Directions

Considerable progress has been made globally to increase the knowledge and understanding of AMR especially in LMIC using effective communication, education, and training. However, due to the high burden of infectious diseases in African countries and the need to lower the incidence of illness through good sanitation, hygiene, and infection prevention measures, African countries should work on:

- Using a continent-wide approach to identify existing gaps in the current regulatory framework for diagnostic tests.
- Promote adoption and enforcement on governance regulations for laboratories in the continent.
- Support the harmonization of a country/continent-led external quality assurance scheme for laboratory management.
- Identify barriers mitigating AMR surveillance and develop pathways of addressing barriers politically, financially, and technically.
- Harness genomic sequencing expertise developed during the COVID-19 pandemic to create a cost-effective snap-shot surveillance system to inform public health approaches.
- Improve the availability of continent-wide AMR data to better understand the prevalence of antimicrobial resistance across human health and animals using the One-Health surveillance approach.

6. Limitation

Some information used for this review was retrieved from TrACSS. The TrACSS assessment used for the AMR surveillance information has several limitations. Firstly, self-reporting bias can arise as countries voluntarily provide data per their assessment. Secondly, inaccurate or incomplete data have been reflected due to incomplete submissions or no submissions by some countries in the region for various reasons, resulting in data gaps. Thirdly, interpretation variability and validity among countries can cause inconsistent responses, complicating data comparison. Therefore, validating the accuracy and reliability of reported data becomes challenging without independent verification. Finally, since the data provided are not associated with any scoring system, it is difficult to interpret the data for comparability between the countries.

7. Conclusions

In conclusion, understanding the full extent and impact of AMR in Africa will not be possible unless we have reliable AMR surveillance and laboratory quality assurance. When available, country data are not consistently gathered, shared, or acknowledged by national entities, which restricts their capacity to have an impact on national actions. Laboratory diagnostics should play a crucial role at every stage of disease management with clinicians relying on it for the diagnosis, monitoring, screening, and prognosis of disease.

Laboratory quality assurance is an important tool for improving the quality of laboratory services and ensuring the reliability of test results. Despite the importance of laboratory quality assurance, many laboratories in Africa face significant challenges in implementing effective quality assurance programs.

AMR surveillance in Africa faces several challenges, including limited laboratory capacity, inadequate data collection and analysis, and poor stakeholder coordination. However, with the growing recognition of AMR's threat, there is increasing momentum to strengthen surveillance systems and improve data sharing and collaboration across the continent which will eventually lead to better control of AMR.

Author Contributions: Conceptualization, I.M.G., K.M. and I.O.; project administration, K.M.; supervision, I.M.G. and J.S.; writing—original draft K.M., I.O., T.A. and A.T.A.; writing—review and editing, I.M.G. and J.S. All authors have read and agreed to the published version of the manuscript.

Funding: This research received no external funding.

Institutional Review Board Statement: Not applicable.

Informed Consent Statement: Not applicable.

Data Availability Statement: No new data were created or analyzed in this study. Data sharing is not applicable to this article.

Acknowledgments: Africa CDC and GLASS for maintaining the database used for some of the analyses.

Conflicts of Interest: The authors declare no conflict of interest.

References

1. Laxminarayan, R.; Duse, A.; Wattal, C.; Zaidi, A.K.M.; Wertheim, H.F.L.; Sumpradit, N.; Vlieghe, E.; Hara, G.L.; Gould, I.M Goossens, H.; et al. Antibiotic resistance—The need for global solutions. *Lancet Infect. Dis.* **2013**, *13*, 1057–1098. [PubMed
2. World Economic Forum. *Global Risk Report 2013*; World Economic Forum: Geneva, Switzerland, 2013.
3. Murray, C.J.L.; Ikuta, K.S.; Sharara, F.; Swetschinski, L.; Aguilar, G.R.; Gray, A.; Han, C.; Bisignano, C.; Rao, P.; Wool, E.; et Global burden of bacterial antimicrobial resistance in 2019: A systematic analysis. *Lancet* **2022**, *399*, 629–655.
4. World Health Organization. *World Health Statistics 2014*; World Health Organization: Geneva, Switzerland, 2014.
5. Okolie, O.J.; Igwe, U.; Ismail, S.U.; Ighodalo, U.L.; Adukwu, E.C. Systematic review of surveillance systems for AMR in Afri *J. Antimicrob. Chemother.* **2023**, *78*, 31–51.
6. Essack, S.; Desta, A.; Abotsi, R.; Agoba, E. Antimicrobial resistance in the WHO African region: Current status and roadmap f action. *J. Public Health* **2017**, *39*, 8–13.
7. World Health Organization. *Antimicrobial Resistance in the WHO African Region: A Systematic Literature Review*; World Heal Organization: Brazzaville, Democratic Republic of the Congo, 2021.
8. Fraser, J.L.; Alimi, Y.H.; Varma, J.K.; Muraya, T.; Kujinga, T.; Carter, V.K.; Schultsz, C.; Del Rio Vilas, V.J. Antimicrobial resistan control efforts in Africa: A survey of the role of Civil Society Organisations. *Glob. Health Action* **2021**, *14*, 1868055.
9. Africa Centres for Disease Control and Prevention (Africa CDC). African Union Commission African Union Framework f Antimicrobial Resistance Control 2020–2025. 2021. Available online: https://africacdc.org/download/african-union-framewor for-antimicrobial-resistance-control-2020-2025/# (accessed on 10 April 2023).
10. Nugent, R.; Back, E.; Beith, A. *The Race against Drug Resistance*; Center for Global Development: Washington, DC, USA, 201 ISBN 978-193328654-9.
11. Iskandar, K.; Molinier, L.; Hallit, S.; Sartelli, M.; Hardcastle, T.C.; Haque, M.; Lugova, H.; Dhingra, S.; Sharma, P.; Islam, S.; et Surveillance of antimicrobial resistance in low- and middle-income countries: A scattered picture. *Antimicrob. Resist. Infe Control.* **2021**, *10*, 63. [CrossRef] [PubMed]

12. O'Neill, J. Tackling Drug-Resistant Infections Globally; Final Report and Recommendations May 2016: Government of the United Kingdom. Available online: http://amr-review.org/sites/default/files/160525_Final%20paper_with%20cover.pdf (accessed on 11 April 2023).
13. Wang, H.; Zhang, W.; Tang, Y.-W. Clinical microbiology in detection and identification of emerging microbial pathogens: Past, present and future. *Emerg. Microbes Infect.* **2022**, *11*, 2579–2589. [PubMed]
14. Tanasiichuk, I.; Karaman, O.; Natrus, L. Key success factors for the implementation of quality management systems in developing countries. *Afr. J. Lab. Med.* **2023**, *12*, 2058. [PubMed]
15. Wilson, M.L.; A Fleming, K.; A Kuti, M.; Looi, L.M.; Lago, N.; Ru, K. Access to pathology and laboratory medicine services: A crucial gap. *Lancet* **2018**, *391*, 1927–1938.
16. Ngo, A.; Gandhi, P.; Miller, W.G. Frequency that laboratory tests influence medical decisions. *J. Appl. Lab. Med.* **2017**, *1*, 410–414.
17. Plebani, M.; Sciacovelli, L. ISO 15189 accreditation: Navigation between quality management and patient safety. *J. Med. Biochem.* **2017**, *36*, 225–230.
18. Tadesse, B.T.; Ashley, E.A.; Ongarello, S.; Havumaki, J.; Wijegoonewardena, M.; González, I.J.; Dittrich, S. Antimicrobial resistance in Africa: A systematic review. *BMC Infect. Dis.* **2017**, *17*, 616.
19. Miller, J.M.; Binnicker, M.J.; Campbell, S.; Carroll, K.C.; Chapin, K.C.; Gilligan, P.H.; Gonzalez, M.D.; Jerris, R.C.; Kehl, S.C.; Patel, R.; et al. A guide to utilization of the microbiology laboratory for diagnosis of infectious diseases: 2018 update by the Infectious Diseases Society of America and the American Society for Microbiology. *Clin. Infect. Dis.* **2018**, *67*, e1–e94. [PubMed]
20. Moyo, K.; Porter, C.; Chilima, B.; Mwenda, R.; Kabue, M.; Zungu, L.; Sarr, A. Use of laboratory test results in patient management by clinicians in Malawi. *Afr. J. Lab. Med.* **2015**, *4*, 8.
21. Iregbu, K.; Osuagwu, C.; Umeokonkwo, C.; Fowotade, A.; Ola-Bello, O.; Nwajiobi-Princewill, P.; Taiwo, S.; Olayinka, A.; Oduyebo, O. Underutilization of the Clinical Microbiology Laboratory by Physicians in Nigeria. *Afr. J. Clin. Exp. Microbiol.* **2020**, *21*, 53.
22. Karaaoui, A.K.E.; Assaf, N. Using the SLIPTA checklist to assess laboratory readiness for Joint Commission International accreditation. *Afr. J. Lab. Med.* **2023**, *12*, 2044.
23. Yao, K.; Luman, E.T.; Nkengasong, J.N.; Maruta, T. The SLMTA programme: Transforming the laboratory landscape in developing countries. *Afr. J. Lab. Med.* **2016**, *5*, 1–8.
24. World Health Organization. *Guide for the Stepwise Laboratory Quality Improvement Process towards Accreditation (SLIPTA) in the WHO African Region—Revision 2*; World Health Organization: Geneva, Switzerland, 2020.
25. World Health Organization. *Global Antimicrobial Resistance Surveillance System (GLASS) Report: Early Implementation 2020*; World Health Organization: Geneva, Switzerland, 2020.
26. Africa CDC Framework for Antimicrobial Resistance—Africa CDC, Africa CDC Framework for Antimicrobial Resistance, 2018–2023. Available online: https://africacdc.org/download/africa-cdc-framework-for-antimicrobial-resistance/ (accessed on 26 April 2023).
27. About the Fleming Fund, The Fleming Fund. Available online: https://www.flemingfund.org/about-us/our-aims/ (accessed on 26 April 2023).
28. Africa Center for Disease Control and Prevention. Antimicrobial Resistance (AMR). Available online: https://africacdc.org/programme/surveillance-disease-intelligence/antimicrobial-resistance-control/ (accessed on 26 April 2023).
29. World Health Organization. AMR Country Progress. Available online: https://amrcountryprogress.org/#/download-responses. (accessed on 26 April 2023).
30. Perovic, O.; Schultsz, C. Stepwise approach for implementation of antimicrobial resistance surveillance in Africa. *Afr. J. Lab. Med.* **2016**, *5*, a482. [CrossRef]
31. Nsubuga, P.; White, M.E.; Thacker, S.B.; Anderson, M.A.; Blount, S.B.; Broome, C.V.; Chiller, T.M.; Espitia, V.; Imtiaz, R.; Sosin, D.; et al. Public Health Surveillance: A Tool for Targeting and Monitoring Interventions. In *Disease Control Priorities in Developing Countries*, 2nd ed.; Jamison, D.T., Breman, J.G., Measham, A.R., Eds.; Chapter 53; Oxford University Press: New York, NY, USA; The International Bank for Reconstruction and Development/The World Bank: Washington, DC, USA, 2006. Available online: https://www.ncbi.nlm.nih.gov/books/NBK11770/ (accessed on 26 April 2023).
32. Kadykalo, S.V.; Anderson, M.E.C.; E Alsop, J. Passive surveillance of antimicrobial resistance in *Salmonella* and *Escherichia coli* isolates from Ontario livestock, 2007–2015. *Am. Jew. Hist.* **2018**, *59*, 617–622.
33. Goddard, L.; Wozniak, T.M. Antimicrobial Resistance Surveillance to Support Decision-Making in a High-Prevalence Region: An Evaluation. *Front. Trop. Dis.* **2021**, *22*, 772491. [CrossRef]
34. Tyson, G.H.; McDermott, P.F.; Li, C.; Chen, Y.; Tadesse, D.A.; Mukherjee, S.; Bodeis-Jones, S.; Kabera, C.; Gaines, S.A.; Loneragan, G.H.; et al. WGS accurately predicts antimicrobial resistance in *Escherichia coli*. *J. Antimicrob. Chemother.* **2015**, *70*, 2763–2769. [CrossRef] [PubMed]
35. Pillay, S.; Calderón-Franco, D.; Urhan, A.; Abeel, T. Metagenomic-based surveillance systems for antibiotic resistance in non-clinical settings. *Front. Microbiol.* **2022**, *13*, 1066995.
36. Grundmann, H.; Gelband, H. Antimicrobial resistance surveillance with whole genome sequencing in Africa: It's (about) time. *Afr. J. Lab. Med.* **2018**, *7*, 3.
37. Siddle, K.J.; Eromon, P.; Barnes, K.G.; Mehta, S.; Oguzie, J.U.; Odia, I.; Schaffner, S.F.; Winnicki, S.M.; Shah, R.R.; Qu, J.; et al. Genomic Analysis of Lassa Virus during an Increase in Cases in Nigeria in 2018. *N. Engl. J. Med.* **2018**, *379*, 1745–1753.

38. Mbala-Kingebeni, P.; Aziza, A.; Di Paola, N.; Wiley, M.R.; Makiala-Mandanda, S.; Caviness, K.; Pratt, C.B.; Ladner, J.T.; Kugelman, J.R.; Prieto, K.; et al. Medical countermeasures during the 2018 Ebola virus disease outbreak in the North Kivu and Ituri Provinces of the Democratic Republic of the Congo: A rapid genomic assessment. *Lancet Infect. Dis.* **2019**, *19*, 648–657. [PubMed]
39. Inzaule, S.C.; Tessema, S.K.; Kebede, Y.; Ouma, A.E.O.; Nkengasong, J.N. Genomic-informed pathogen surveillance in Africa: Opportunities and challenges. *Lancet Infect. Dis.* **2021**, *21*, e281–e289. [PubMed]
40. Framework for One Health Practice in National Public Health Institutes—Africa CDC. Available online: https://africacdc.org/download/framework-for-one-health-practice-in-national-public-health-institutes/ (accessed on 26 April 2023).

antibiotics MDPI

ystematic Review

'redicting Extended-Spectrum Beta-Lactamase .nd Carbapenem Resistance in Enterobacteriaceae Bacteremia: \ Diagnostic Model Systematic Review and Meta-Analysis

ristan T. Timbrook [1,2,*] and McKenna J. Fowler [1]

[1] Department of Pharmacotherapy, University of Utah College of Pharmacy, Salt Lake City, UT 84112, USA; mckenna.fowler@pharm.utah.edu
[2] BioMérieux, 69280 Marcy l'Etoile, France
* Correspondence: tristan.timbrook@utah.edu

Abstract: Enterobacteriaceae bacteremia, particularly when associated with antimicrobial resistance, can result in increased mortality, emphasizing the need for timely effective therapy. Clinical risk prediction models are promising tools, stratifying patients based on their risk of resistance due to ESBL and carbapenemase-producing Enterobacteriaceae in bloodstream infections (BSIs) and, thereby, improving therapeutic decisions. This systematic review and meta-analysis synthesized the literature on the performance of these models. Searches of PubMed and EMBASE led to the identification of 10 relevant studies with 6106 unique patient encounters. Nine studies concerned ESBL prediction, and one focused on the prediction of carbapenemases. For the two ESBL model derivation studies, the discrimination performance showed sensitivities of 53–85% and specificities of 93–95%. Among the four ESBL model derivation and validation studies, the sensitivities were 43–88%, and the specificities were 77–99%. The sensitivity and specificity for the subsequent external validation studies were 7–37% and 88–96%, respectively. For the three external validation studies, only two models were evaluated across multiple studies, with a pooled AUROC of 65–71%, with one study omitting the sensitivity/specificity. Only two studies measured clinical utility through hypothetical therapy assessments. Given the limited evidence on their interventional application, it would be beneficial to further assess these or future models, to better understand their clinical utility and ensure their safe and impactful implementation.

Keywords: Enterobacteriaceae bacteremia; antimicrobial resistance; ESBL production; carbapenemase production; risk prediction scoring systems; clinical utility of predictive models; antimicrobial stewardship

1. Introduction

More than 2.8 million antibiotic-resistant infections occur in the United States each year, and more than 35,000 people die as a result [1,2]. Resistant pathogens of major global health concern on the CDC's urgent and serious threat lists include carbapenem-resistant Enterobacteriaceae (CRE) and extended-spectrum-beta-lactamase-producing Enterobacteriaceae (ESBL-E) [1]. The use of broad-spectrum antibiotics has resulted in the development of resistance in many bacterial pathogens [3]. Thus, the use of broad-spectrum agents, such as carbapenems, is often avoided when possible [4]. This conundrum of ineffective antimicrobial medications and resistant microorganisms places a burden on patients and healthcare providers [5]. Identifying the risk factors for resistant Gram-negative infections is imperative in guiding appropriate treatment interventions and reducing the utilization of broad-spectrum antibiotics [6]. Moreover, employing these initiatives to accurately predict antimicrobial resistance and assign targeted effective treatment is crucial to the recovery of patients with resistant Gram-negative bloodstream infections (GN-BSIs) [7].

Currently, published models exist that assess clinical risk factors for predicting whether a patient has a resistant Gram-negative bloodstream infection [8]. These models often assess factors such as hospitalization, previous colonization, previous antibiotic therapy, age, catheter use, indwelling hardware, surgery, etc. [6]. The application of prediction models can support antimicrobial stewardship efforts by aiding the selection of appropriate treatments and reducing the selection pressure from antibiotics that leads to resistant pathogens [6]. The prediction of resistance in these organisms can be enhanced through evaluation of the performance of published prediction models. This could enhance the understanding of the clinical performance of these models, and their utility in healthcare institutions for effective clinical decision-making. A review of the risk factors and scoring systems used to predict ESBL-E and CRE infections was recently published [8]. However, systematic evaluations and the potential pooled performance of such tools are lacking.

Given the increasing threat from antibiotic-resistant infections, and the gaps in the literature regarding the prediction of ESBL-E and CRE bloodstream infections, this diagnostic model systematic review aims to evaluate the performance of the available models in the literature. This could inform healthcare institutions' decision-making processes via evaluation of the integration of such models into their clinical procedures to reduce healthcare costs and improve patient outcomes. Moreover, this study could facilitate the adoption of high-performing models to further enhance effective clinical decision-making and antimicrobial stewardship efforts. This diagnostic model systematic review aims to synthesize the available literature to evaluate the performance and discriminatory capabilities of the published models used in predicting ESBL and carbapenemase-producing Enterobacteriaceae bloodstream infections.

2. Results

2.1. Study Selection

The literature search resulted in 2279 studies that met our keyword query (Figure 1). After the removal of duplicates, we searched 1811 studies, and this resulted in five studies identified for inclusion. Searching the bibliographies of the included studies identified an additional four studies for inclusion. One additional study was included from a conference proceeding. In total, data were extracted from 10 studies for analysis, with 6106 unique patient encounters [9–18].

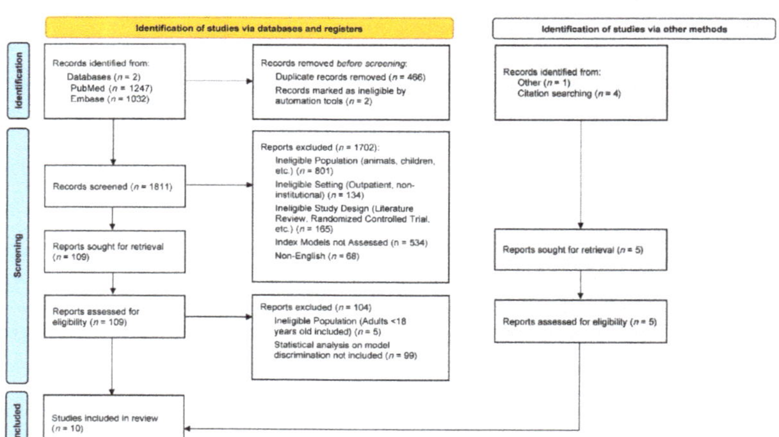

Figure 1. The study selection procedure.

2.2. Study and Model Characteristics

The characteristics of the studies included can be found in Table 1. The studies were primarily conducted in the United States (70%). Nearly all the studies had a retrospective cohort design (80%), with two retrospective case–control designs, and no prospective studies. The sample sizes varied, ranging from 145 to 1288 patients. The majority of studies

(90%) were ESBL models with a resistance incidence ranging from 5.7% to 33.3%, and a median of 12.7%. A single study was of a CRE model, and did not report resistance incidence, as it was a case–control study. Among the 10 studies, 19 analyses of models were evaluated, with four (21.1%) being derivation-based, five (26.3%) being derivation- and internal-validation-based, and ten (52.6%) being external-validation-based.

Table 1. Study characteristics.

First Author, Year	Country	Design	Model Type	Sample Size	Resistance Type	Resistance Incidence
Andrews 2023 [9]	USA	Retrospective cohort	External validation of Lee 2017 and Augustine 2017	356	ESBL	11.5%
Augustine 2017 [10]	USA	Retrospective cohort	Derivation and internal validation	910	ESBL	4.6%
Cwengros 2020 [11]	USA	Retrospective cohort	External validation of Lee 2017 and Augustine 2017	451	ESBL	16%
Goodman 2016 [12]	USA	Retrospective cohort	Derivation and internal validation	1288	ESBL	15%
Goodman 2019 [13]	USA	Retrospective cohort	Derivation and internal validation	1288	ESBL	15%
Holmgren 2020 [14]	Sweden	Retrospective cohort	Derivation	625	ESBL	9%
Lee 2017 [15]	Taiwan	Retrospective cohort	Derivation	1141	ESBL	5.7%
Madrid-Morales 2021 [16]	USA	Retrospective cohort	External validation of Augustine, 2017; Goodman, 2016 and 2019; and Tumbarello, 2011	145	ESBL	13.8%
Tumbarello 2011 [17]	Italy	Retrospective case–control	Derivation and external validation	339; 510	ESBL	33.3%; 20%
Weston 2020 [18]	USA	Retrospective case–control	Derivation and internal validation	341	CRE	NA

The demographic details across the studies were heterogeneously reported. Patients predominantly had a median or mean age in the mid-60s, with the female representation ranging from 45.4% to 56.9% (Supplementary Table S1). The racial and ethnic distributions demonstrated substantial diversity, with African American individuals constituting up to 72.5% of patients in some cohorts, while, in others, White individuals comprised up to 47.2%, though the majority of the studies did not report race or ethnicity. Diabetes was consistently highlighted as a significant comorbidity, with prevalence rates as high as 40.1%. Other notable comorbidities included congestive heart failure (up to 50.2%) and end-stage renal disease (23.9%). The source of infection was consistently highest for urinary tract infections, ranging from 31.7% to 58.6%. Although the organisms evaluated varied across studies, *E. coli* was the dominant causative organism, observed in proportions ranging from 56% to 77.2%, followed by *K. pneumoniae*, with figures fluctuating between 13.5% and 40%, except for the single carbapenem-resistance-prediction study, which focused solely on *K. pneumoniae*.

Though the models displayed variations in the risk factors for resistance, certain factors consistently emerged across studies (Table 2). Regarding demographics, age was a predictor in two studies. Hospitalization or admission histories were also common predictors, including prior hospitalizations, recent emergency department visits, and having received care in high-burden regions or abroad. A variety of medical histories were frequently used as predictors, including ESBL-E infection or colonization, in addition to comorbidities or medical exposures, such as COPD, emphysema, ventilator dependence, and urological diseases, and the presence of chronic indwelling hardware. Procedures or interventions were predictors in two studies, and included invasive procedures and GI/GU procedures. Medication-use history was a consistent predictor with antibiotic exposure, both in terms of the duration (days or weeks), and with specific mentions of antibiotic classes, such as β-lactams and fluoroquinolones.

Table 2. Model characteristics.

Study	Demographics	Hospitalization/ Admission History	Medical History	Procedure/ Intervention History	Medication Use History	Other
Augustine 2017 [10]	-	-	ESBL-E infection or colonization	GI/GU procedure	Number of prior BL/FQ courses	-
Goodman 2016 [12]	Age, ≥43 years	Hospitalization in ESBL high-burden region	ESBL-E infection or colonization, chronic indwelling vascular hardware	-	Days of antibiotic exposure	
Goodman 2019 [13]	-	Hospital care abroad, prior hospitalization	COPD, emphysema, ventilator dependence, indwelling hardware, MDRO colonization or infection	-	Weeks of antibiotic exposure	Source of infection
Hömgren 2019 [14]	-	Hospital care abroad	ESBL-E infection or colonization	-	-	-
Lee 2017 [15]	-	Prior hospitalization, recent ED visits	Urological diseases, diabetes mellitus	Invasive procedure	Antibiotic exposure	Nursing home residents
Tumbarello 2011 [17]	Age, ≥70 years	Recent hospitalization, admission from healthcare	Charlson comorbidity index, urinary catheterization	-	Previous therapy with β-lactams and/or fluoroquinolones	-
Weston 2020 [18]	-	Admitted >3 days	Prior CRE culture, liver disease, mechanical ventilation	-	Proton pump inhibitor, antibiotic exposure	Admission from SNF, no prior culture

NOTE. BL: β-lactams; COPD: chronic obstructive pulmonary disease; CRE: carbapenem-resistant Enterobacteriaceae; ED: emergency department; ESBL-E: extended-spectrum β-lactamase–producing Enterobacteriaceae; FQ: fluoroquinolones; GI: gastrointestinal; GU: genitourinary; MDRO: multidrug-resistant organisms; SNF: skilled nursing facility.

2.3. Results of Model Performance in Individual and Pooled Studies

Where reported, the discrimination performance based on the sensitivity and specificity of derivation, internal validation, and external validation studies for ESBL models can be found in Figure 2. The derivation performance ranged in sensitivity from 53 to 85%, and in specificity from 93 to 95%, with medians of 55% and 94%, respectively. The derivation and internal validation performance ranged from 43 to 88% for sensitivity and 77 to 99% for specificity, with medians of 43% and 90%, respectively. With the exception a single study that performed derivation and an external validation (Tumbarello; external validation sensitivity 73%, and specificity 95%), subsequent studies performing external validations had a performance that ranged from 7 to 37% for sensitivity, and from 88 to 96% for specificity, with medians of 32% and 90%, respectively. One external validation study did not report the sensitivity and specificity performances, but did report the AUROCs for the Augustine, Goodman 2016, Goodman 2019, and Tumbarello models, which were 0.71 (95% CI 0.65–0.78), 0.59 (95% CI 0.52–0.66), 0.74 (95% CI 0.67–0.81), and 0.76 (95% CI 0.69–0.82), respectively. Two models were externally evaluated in more than one study, and had a pooled discrimination performance, with AUROCs of 0.71 (95% CI 0.66 to 0.75) and 0.65 (95% CI 0.56 to 0.73) (Figure 3). Only three studies reported calibration evaluations, all reporting an adequate fit either from calibration plots, Hosmer–Lemeshow testing, or both. No studies published their model equations.

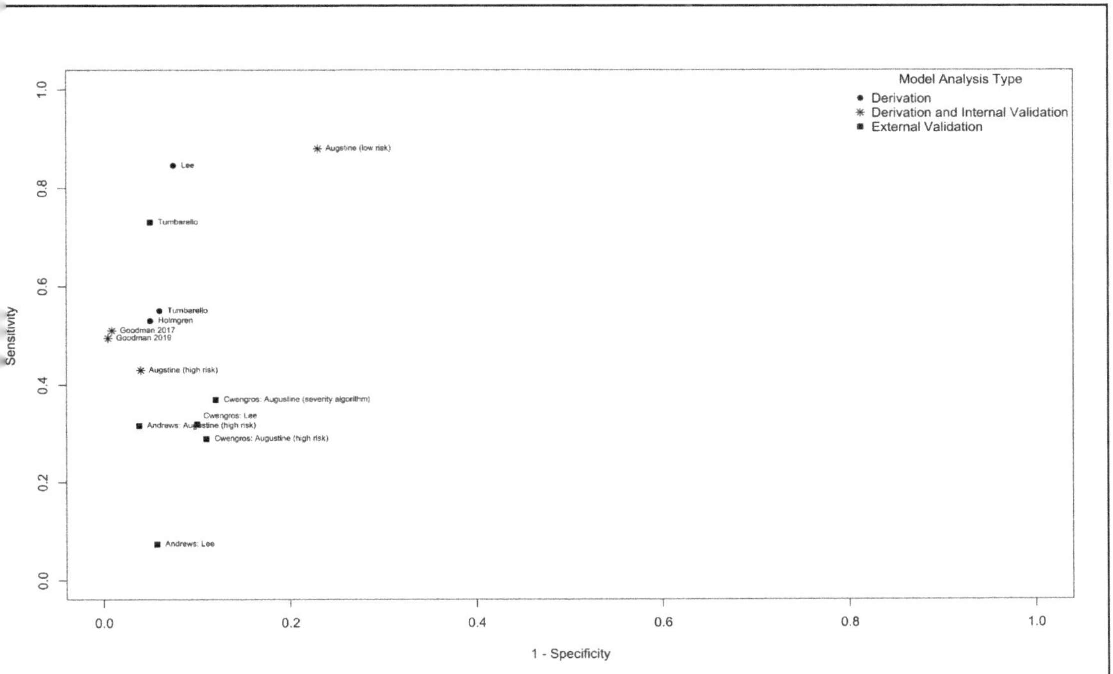

Figure 2. The discrimination performance of ESBL prediction models studies.

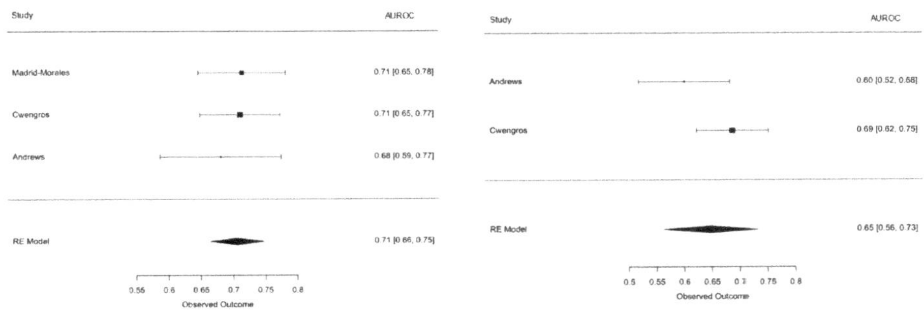

(**a**) (**b**)

Figure 3. The pooled discrimination performance of external validations. (**a**) Augustine 2017 [10] ESBL model. (**b**) Lee 2017 [15] ESBL model.

2.4. Clinical Utility

Only two studies (20%) evaluated the potential therapy impacts of model use. In Cwengros et al., using the Lee model, 10% were overtreated with ceftriaxone-susceptible isolates, and 69% were undertreated for resistant ones. Using a severity-of-illness-stratified Augustine model, Cwengros et al. reflected 12% overtreated, and 63% undertreated. Finally, using the Augustine model without severity-of-illness stratification. Cwengros et al. showed results of 11% overtreated and 71% undertreated. Similarly, Andrews et al. observed that 6% of patients with susceptible isolates were overtreated via the Lee model, and that 93% of resistant isolates were undertreated. Using the Augustine model without severity-of-illness stratification, Andrews et al. noted results of 4% overtreated and

68% undertreated. Both the study by Cwengros and that by Andrews et al. evaluated the total carbapenem consumption per 1000 patient days, with similar results.

3. Discussion

Due to the increasing burden of antimicrobial resistant infections, tools are needed to improve appropriate antimicrobial use, and achieve the antimicrobial stewardship goals of providing the right drug to the right patient at the right time. In our systematic review of the literature, we found 10 studies with 19 analyses of various models for ESBL and carbapenem resistance among Enterobacteriaceae bloodstream infections. While ESBL derivation studies noted a moderate performance, with the median sensitivity and specificity of 55% and 94%, subsequent studies with external validations reflected a lower median sensitivity and specificity of 32% and 90%, respectively. Two ESBL models with multiple external validations showed a pooled performance with AUROCs of 65% and 71%. Clinical utility evaluations were limited to two studies that reflected a low overtreatment in ceftriaxone-susceptible isolates with the use of models, but a substantial (>60%) undertreatment in ceftriaxone-resistant ones. Given these findings, while some models may limit overtreatment, there is still a critical need to refine and improve current models, to address the substantial risk of undertreatment, ensuring optimal patient care, and advancing antimicrobial stewardship goals.

Endorsed by Infectious Diseases Society of America guidelines, the local validation of risk prediction models, and thorough reporting, are paramount to understanding their performance and potential clinical utility [19]. In our review, less than half of the studies were external validations. Notably, where these validations were performed, they demonstrated an inferior performance compared to the original studies—a decrease which is often seen when models undergo external validation [20].

Our understanding of a model's generalizability and reproducibility is facilitated by external validation [21]. Specifically, this process can clarify the model's discrimination (the ability to correctly classify patients into risk groups) and calibration (the alignment between the observed vs. predicted risks across predicted risk ranges) [22]. A significant limitation we observed was that none of the derivation models disclosed their equations, impeding their use for external calibration assessments by frontline clinicians. These studies opted to transform the equations into risk scores for easier external application. Yet, this omission also prohibits the often necessary re-calibration of the risk model's intercept for differences in baseline risk, particularly when considering the observed heterogeneity in ESBL rates. Given such variations in baseline risk and predictor performance, the external application of models will generally require a model rebuild, rather than reliance on risk scores.

Only three of the studies we analyzed reported calibration performance evaluations, a key metric that is crucial to understanding how prediction models perform across diverse risk levels. The nuances of the demographics and health backgrounds in these studies present a significant variable. For instance, despite differing demographics, such as the elderly veteran population, in the Madrid-Morales study, its discrimination performance was unexpectedly comparable to the Cwengros study. Such similarities in discrimination, despite potential age-related susceptibilities in predicting resistance or varying comorbidities such as COPD, suggest that, while the discrimination might align, the calibration could vary significantly, due to differences in the distribution of predictors across studies. In essence, two models might classify risk similarly (discrimination), but could differ in how accurately those risk predictions match the actual outcomes (calibration) when applied to different populations. This underlines the critical importance of calibration evaluations, which provide insight into how well predictions align with actual outcomes across different risk strata and varied predictor distributions. Given these observations, the comprehensive reporting of calibration evaluations becomes paramount for ensuring the robust application of prediction models across diverse settings and populations.

In our review, only two of the reviewed studies reported methods of sample-size consideration for their prediction model, and these indicated a planned adherence to historical minimal recommendations of 10 events per variable (EPV) for modeling sample size requirements [23]. Surprisingly, no study undertook formal sample-size calculations for prediction modeling, and one study did not meet the minimal EPV recommendations. Given the elements noted above regarding the overall methodological approach of current studies, we strongly advocate for future studies to adhere to the Transparent Reporting of a Multivariable Prediction Model for Individual Prognosis or Diagnosis (TRIPOD) Reporting Guidelines. Additionally, it is imperative that future studies employ the available tools for sample size evaluations, and consistently evaluate and report the calibration performance of their models [24–26].

Clinical utility evaluations of predictive models are important in understanding the utility of a model beyond its discrimination and calibration performance; however, few studies in our review evaluated clinical utility [27]. As noted, in the two studies evaluating clinical utility, only the hypothetical undertreatment, overtreatment, and carbapenem use per 1000 patient days was evaluated. However, the collaborators of the Augustine study recently published a separate subsequent study evaluating the implementation of their risk prediction scoring on patient management [28]. For their implementation, they made the prediction score and algorithm available in print, on their internal website for providers, and through a mobile app. They noted a significant decrease in the time taken to administer appropriate antimicrobial therapy from 78 h to 46 h after the implementation of the risk prediction score ($p = 0.04$). This study offers a promising testament to the potential of locally derived, internally validated, and clinically implemented ESBL bloodstream infection risk scoring tools.

However, the landscape of resistance prediction is not confined to these tools alone. Of particular note, genotypic prediction of resistance from diagnostic testing has been shown to outperform some ESBL risk-prediction models [11]. There are myriad factors to weigh when selecting a tool for clinical utility and local validation, as delineated elsewhere [8]. In light of such advancements, institutions should remain adaptive, continuously evaluating a broad spectrum of tools. These evaluations should align with their unique patient demographics and institutional needs, and the evolving epidemiology of resistance [29,30].

An additional important consideration in our review is the inclusion of only a single study focusing on predicting carbapenem resistance in Enterobacteriaceae bloodstream infections. The limited representation of CRE is noteworthy for several reasons. Firstly, CRE is an extremely clinically challenging infection to manage, associated with significant morbidity and mortality [31]. Secondly, CRE detections in bloodstream infections are much more infrequent in the United States, where the majority (70%) of the models included in our systematic review were developed, possibly reflecting differences in clinical practice and research priorities versus countries with higher detection rates. Finally, the singular representation of CRE in our results may reflect a publication bias or a gap in the existing research that warrants future studies. However, despite its limited representation, the inclusion of the CRE prediction model in Enterobacteriaceae bloodstream infections serves to highlight the current state of the literature on prediction model tools to support patient management and antimicrobial stewardship, along with the need for additional research in this area.

There are several limitations to our review. The selection of studies was limited to studies evaluating risk modeling for ESBL and carbapenem resistance in Enterobacteriaceae in the bloodstream, which may limit practical applications, as it requires a positive blood culture and rapid organism identification method, to determine eligibility for the tool and provide risk calculation before automated susceptibility testing, in order to appropriately benefit from these approaches [32,33]. Models have been developed elsewhere for general resistance to empiric therapy for both sepsis and bloodstream infections, and future diagnostic prediction model systematic reviews should evaluate the performance of available models for these clinical scenarios, to inform clinical practice [34,35]. As noted previously in the discussion, quality indicators for performance, including calibration and the clinical utility of models, were missing from the majority of studies, limiting the interpretability of the performance and useability of the models [36–38]. Additionally, the external validation studies were limited, further challenging the evaluation of model performance. Future studies should focus on the external validation of existing models and clinical utility determinations. Similarly, only 30% of the studies took place outside of the United States, which makes the generalizability or transportability of such approaches and related performance to non-US countries unclear [39]. Finally, our study was limited to only grey literature (e.g., conference proceedings) for two leading infectious disease conferences, which may have led to the omission of relevant research from other conferences, particularly outside of infectious diseases.

4. Materials and Methods

This systematic review and meta-analysis was reported in accordance with the Transparent Reporting of Multivariable Prediction Models for Individual Prognosis or Diagnosis: Checklist for Systematic Reviews and Meta-Analyses (TRIPOD-SRMA) (see Supplementary Table S2) [40]. This study was not preregistered.

4.1. Study Selection

We performed a literature search using PubMed and EMBASE, using the time period from each database's inception to 3 September 2023, and including English-language studies. A search strategy for keywords related to Gram-negative AMR and prediction models was employed (Supplementary Table S3). In addition to searching these databases, we hand-searched infectious disease major conference proceedings (ECCMID and IDWeek) related to our topic for the last two years, covering the period of 2022–2023, to capture early-stage data that might still be pending full publication.

The eligible studies included cohort and case–control observational studies evaluating a population of adult patients ($\geq$18 years old) with Enterobacterales BSI and risk prediction models, including risk scores. Both the derivation and validation model types were eligible for inclusion. The outcome of the risk-prediction models was AMR including ESBL and CRE. The timing for the model included models used at the suspected onset of bloodstream infection for both community-onset and healthcare-acquired infection. The setting of the models was for use in hospital-admitted or emergency department patients. Studies with multivariable models evaluating the association of predictors with AMR, but which did not attempt to develop the model for prediction (e.g., the predictive performance of the model was not evaluated), were excluded.

Two reviewers (M.F. and T.T.T.) independently assessed the identified records using the eligibility criteria above. Duplicates were initially removed, before articles were reviewed. A first-pass review of titles and abstracts was performed, with selection disagreements resolved through consensus. A second-pass review of full-text articles was performed, with the additional step of resolving final discordances though consensus. A bibliography review of the included articles (i.e., snowballing) was performed to identify additional articles for inclusion.

4.2. Data Extraction

Extracted data from the final selection of articles were included in the review into a standardized Microsoft Excel grid, based on guidance from the CHARMS checklist [41]. The PICO framework (population, index model, comparator model, outcome[s]) information was extracted for each study, using a standardized form [42]. Additionally, the source of the data, the sample size, the number of participants with AMR, the handling of missing data, the selection of predictors, the predictive performance (the overall model fit, discrimination, and calibration including standard errors or confidence intervals) and, where present, measures of clinical utility (e.g., net benefit, decision curve analysis) were reviewed. No authors were contacted for missing data.

4.3. Statistical Analysis

The study characteristics, including the model performance, were reported with descriptive statistics, or through figures. Meta-analysis using random effects modeling was performed for models where a specific model had been externally validated [43]. All the data analyses, including the meta-analyses, were performed using R version 4.0.1 and 'metafor' for analysis [44].

5. Conclusions

In conclusion, our review reveals that a multitude of resistance-in-Enterobacteriaceae bloodstream prediction models exist in the literature, with a minority of studies concerning the external validation of existing models. The models showed a moderate discrimination performance, with a lower performance in external validations. A local external validation of the performance of the model discrimination and calibration is needed in order to understand the use of these models in specific settings, along with potential clinical utility evaluations versus the baseline appropriate prescription in Enterobacteriaceae bacteremia. Due to the scarcity of reports on the interventional use of these models, further clinical evaluation is required, to determine their clinical utility and safety. Evaluating these or newer models will strengthen the assessment of the efficacy of such approaches and their impact on patient outcomes, thus informing healthcare institutions when it comes to potentially improving therapy management, reducing healthcare costs, and enhancing patient outcomes in Enterobacteriaceae bloodstream infections.

Supplementary Materials: The following supporting information can be downloaded at: https://www.mdpi.com/article/10.3390/antibiotics12091452/s1, Table S1: Patient characteristics; Table S2: TRIPOD-SRMA Checklist for Reporting Systematic Reviews of Prediction Model Studies; Table S3: Keyword search.

Author Contributions: Conceptualization, T.T.T. and M.J.F.; methodology, T.T.T. and M.J.F.; formal analysis, T.T.T.; writing—original draft preparation, T.T.T.; writing—review and editing, T.T.T. and M.J.F. All authors have read and agreed to the published version of the manuscript.

Funding: This research received no external funding.

Data Availability Statement: The data used for all analyses and analytic code are available from the corresponding author upon reasonable request.

Conflicts of Interest: T.T.T. is an employee and shareholder of bioMérieux, an in vitro diagnostics company.

eferences

CDC. *Antibiotic Resistance Threats in the United States*; U.S. Department of Health and Human Services: Atlanta, GA, USA, 2019.

van Duin, D.; Paterson, D.L. Multidrug-Resistant Bacteria in the Community: An Update. *Infect. Dis. Clin. N. Am.* **2020**, *34*, 709–722. [CrossRef]

Serwecińska, L. Antimicrobials and Antibiotic-Resistant Bacteria: A Risk to the Environment and to Public Health. *Water* **2020**, *12*, 3313. [CrossRef]

Cizman, M.; Plankar Srovin, T. Antibiotic consumption and resistance of gram-negative pathogens (collateral damage). *GMS Infect. Dis.* **2018**, *6*, Doc05. [CrossRef]

Smith, R.A.; M'Ikanatha N, M.; Read, A.F. Antibiotic resistance: A primer and call to action. *Health Commun.* **2015**, *30*, 309–314. [CrossRef] [PubMed]

Palacios-Baena, Z.R.; Giannella, M.; Manissero, D.; Rodríguez-Baño, J.; Viale, P.; Lopes, S.; Wilson, K.; McCool, R.; Longshaw, C. Risk factors for carbapenem-resistant Gram-negative bacterial infections: A systematic review. *Clin. Microbiol. Infect.* **2021**, *27*, 228–235. [CrossRef] [PubMed]

Lodise, T.P.; Zhao, Q.; Fahrbach, K.; Gillard, P.J.; Martin, A. A systematic review of the association between delayed appropriate therapy and mortality among patients hospitalized with infections due to Klebsiella pneumoniae or Escherichia coli: How long is too long? *BMC Infect. Dis.* **2018**, *18*, 625. [CrossRef]

Li, C.; Claeys, K.C.; Justo, J.A.; Heil, E.L. Heil. No Crystal Ball? Using Risk Factors and Scoring Systems ot Predict Extended-Spectrum Beta-Lactamase Producing Enterobacterales (ESBL-E) and Carbapenem-Resistant Enterobacterales (CRE) Infections. *Curr. Infect. Dis. Rep.* **2022**, *24*, 147–158. [CrossRef]

9. Andrews, S.R.; Timbrook, T.T.; Fisher, M.A.; Tritle, B.J. Evaluation of a Rapid Diagnostic Test and Scoring Tools for Predicti Ceftriaxone Resistance in Enterobacterales Bloodstream Infections. In Proceedings of the IDWeek 2023, Boston, MA, USA, 11– October 2023.
10. Augustine, M.R.; Testerman, T.L.; Justo, J.A.; Bookstaver, P.B.; Kohn, J.; Albrecht, H.; Al-Hasan, M.N. Clinical Risk Score f Prediction of Extended-Spectrum β-Lactamase-Producing Enterobacteriaceae in Bloodstream Isolates. *Infect. Control Ho Epidemiol.* **2017**, *38*, 266–272. [CrossRef]
11. Cwengros, L.N.; Mynatt, R.P.; Timbrook, T.T.; Mitchell, R.; Salimnia, H.; Lephart, P.; Pogue, J.M. Minimizing Time to Optim Antimicrobial Therapy for Enterobacteriaceae Bloodstream Infections: A Retrospective, Hypothetical Application of Predicti Scoring Tools vs Rapid Diagnostics Tests. *Open Forum Infect. Dis.* **2020**, *7*, ofaa278. [CrossRef]
12. Goodman, K.E.; Lessler, J.; Cosgrove, S.E.; Harris, A.D.; Lautenbach, E.; Han, J.H.; Milstone, A.M.; Massey, C.J.; Tamma, P. Antibacterial Resistance Leadership, G. A Clinical Decision Tree to Predict Whether a Bacteremic Patient Is Infected with Extended-Spectrum β-Lactamase-Producing Organism. *Clin. Infect. Dis.* **2016**, *63*, 896–903. [CrossRef]
13. Goodman, K.E.; Lessler, J.; Harris, A.D.; Milstone, A.M.; Tamma, P.D. A methodological comparison of risk scores versus decisi trees for predicting drug-resistant infections: A case study using extended-spectrum beta-lactamase (ESBL) bacteremia. *Infe Control Hosp. Epidemiol.* **2019**, *40*, 400–407. [CrossRef] [PubMed]
14. Holmgren, A.; Ljung, A.; Bremell, D. An easy-to-use scoring system for predicting bacteraemia with third-generati cephalosporin-resistant Enterobacterales in a low-resistance setting. *Infect. Dis.* **2020**, *52*, 242–248. [CrossRef]
15. Lee, C.H.; Chu, F.Y.; Hsieh, C.C.; Hong, M.Y.; Chi, C.H.; Ko, W.C.; Lee, C.C. A simple scoring algorithm predicting extende spectrum β-lactamase producers in adults with community-onset monomicrobial Enterobacteriaceae bacteremia: Matters frequent emergency department users. *Medicine* **2017**, *96*, e6648. [CrossRef]
16. Madrid-Morales, J.; Sharma, A.; Reveles, K.; Velez-Mejia, C.; Hopkins, T.; Yang, L.; Walter, E.; Cadena, J. Validation of Availab Extended-Spectrum-Beta-Lactamase Clinical Scoring Models in Predicting Drug Resistance in Patients with Enteric Gra Negative Bacteremia Treated at South Texas Veterans Health Care System. *Antimicrob. Agents Chemother.* **2021**, *65*. [CrossR [PubMed]
17. Tumbarello, M.; Trecarichi, E.M.; Bassetti, M.; De Rosa, F.G.; Spanu, T.; Di Meco, E.; Losito, A.R.; Parisini, A.; Pagani, Cauda, R. Identifying patients harboring extended-spectrum-beta-lactamase-producing Enterobacteriaceae on hospital admissi Derivation and validation of a scoring system. *Antimicrob. Agents Chemother.* **2011**, *55*, 3485–3490. [CrossRef]
18. Weston, G.; Jahufar, F.; Sharma, N.; Su, C.; Bellin, E.; Ostrowsky, B. Derivation of a Model to Guide Empiric Therapy f Carbapenem-Resistant Klebsiella pneumoniae Bloodstream Infection in an Endemic Area. *Open Forum Infect. Dis.* **2020**, *7*, ofaa0 [CrossRef] [PubMed]
19. Metlay, J.P.; Waterer, G.W.; Long, A.C.; Anzueto, A.; Brozek, J.; Crothers, K.; Cooley, L.A.; Dean, N.C.; Fine, M.J.; Flanders, S. et al. Diagnosis and Treatment of Adults with Community-acquired Pneumonia. An Official Clinical Practice Guideline the American Thoracic Society and Infectious Diseases Society of America. *Am. J. Respir. Crit. Care Med.* **2019**, *200*, e45–e [CrossRef]
20. Siontis, G.C.; Tzoulaki, I.; Castaldi, P.J.; Ioannidis, J.P. External validation of new risk prediction models is infrequent and reve worse prognostic discrimination. *J. Clin. Epidemiol.* **2015**, *68*, 25–34. [CrossRef] [PubMed]
21. Debray, T.P.; Vergouwe, Y.; Koffijberg, H.; Nieboer, D.; Steyerberg, E.W.; Moons, K.G. A new framework to enhance t interpretation of external validation studies of clinical prediction models. *J. Clin. Epidemiol.* **2015**, *68*, 279–289. [CrossRef]
22. Ramspek, C.L.; Jager, K.J.; Dekker, F.W.; Zoccali, C.; van Diepen, M. External validation of prognostic models: What, why, ho when and where? *Clin. Kidney J.* **2021**, *14*, 49–58. [CrossRef]
23. Peduzzi, P.; Concato, J.; Kemper, E.; Holford, T.R.; Feinstein, A.R. A simulation study of the number of events per variable logistic regression analysis. *J. Clin. Epidemiol.* **1996**, *49*, 1373–1379. [CrossRef] [PubMed]
24. Collins, G.S.; Reitsma, J.B.; Altman, D.G.; Moons, K.G. Transparent reporting of a multivariable prediction model for Individu Prognosis or Diagnosis (TRIPOD): The TRIPOD statement. *J. Clin. Epidemiol.* **2015**, *68*, 134–143. [CrossRef]
25. Pate, A.; Riley, R.D.; Collins, G.S.; van Smeden, M.; Van Calster, B.; Ensor, J.; Martin, G.P. Minimum sample size for developi a multivariable prediction model using multinomial logistic regression. *Stat. Methods Med. Res.* **2023**, *32*, 555–571. [CrossR [PubMed]
26. Snell, K.I.E.; Archer, L.; Ensor, J.; Bonnett, L.J.; Debray, T.P.A.; Phillips, B.; Collins, G.S.; Riley, R.D. External validation of clinic prediction models: Simulation-based sample size calculations were more reliable than rules-of-thumb. *J. Clin. Epidemiol.* **20** *135*, 79–89. [CrossRef]
27. Steyerberg, E.W.; Vergouwe, Y. Towards better clinical prediction models: Seven steps for development and an ABCD f validation. *Eur. Heart J.* **2014**, *35*, 1925–1931. [CrossRef] [PubMed]
28. Haimerl, B.J.; Encinas, R.; Justo, J.A.; Kohn, J.; Bookstaver, P.B.; Winders, H.R.; Al-Hasan, M.N. Optimization of Empiric Antimicrobial Therapy in Enterobacterales Bloodstream Infection Using the Extended-Spectrum Beta-Lactamase Prediction Sco *Antibiotics* **2023**, *12*, 1003. [CrossRef] [PubMed]
29. Pogue, J.M.; Heil, E.L. The Clinical Impact of a Negative Molecular beta-Lactamase Gene Test for Enterobacteriaceae: Let's N Let Perfect Be the Enemy of Really Good. *J. Clin. Microbiol.* **2020**, *58*. [CrossRef] [PubMed]

30. Pogue, J.M.; Heil, E.L.; Lephart, P.; Johnson, J.K.; Mynatt, R.P.; Salimnia, H.; Claeys, K.C. An Antibiotic Stewardship Program Blueprint for Optimizing Verigene BC-GN within an Institution: A Tale of Two Cities. *Antimicrob. Agents Chemother.* **2018**, *62*. [CrossRef]
31. Tamma, P.D.; Aitken, S.L.; Bonomo, R.A.; Mathers, A.J.; van Duin, D.; Clancy, C.J. Infectious Diseases Society of America 2023 Guidance on the Treatment of Antimicrobial Resistant Gram-Negative Infections. *Clin. Infect. Dis.* **2023**, ciad428. [CrossRef]
32. Lagace-Wiens, P.R.; Adam, H.J.; Karlowsky, J.A.; Nichol, K.A.; Pang, P.F.; Guenther, J.; Webb, A.A.; Miller, C.; Alfa, M.J. Identification of blood culture isolates directly from positive blood cultures by use of matrix-assisted laser desorption ionization-time of flight mass spectrometry and a commercial extraction system: Analysis of performance, cost, and turnaround time. *J. Clin. Microbiol.* **2012**, *50*, 3324–3328. [CrossRef]
33. Idelevich, E.A.; Schule, I.; Grunastel, B.; Wullenweber, J.; Peters, G.; Becker, K. Rapid identification of microorganisms from positive blood cultures by MALDI-TOF mass spectrometry subsequent to very short-term incubation on solid medium. *Clin. Microbiol. Infect.* **2014**, *20*, 1001–1006. [CrossRef] [PubMed]
34. MacFadden, D.R.; Coburn, B.; Shah, N.; Robicsek, A.; Savage, R.; Elligsen, M.; Daneman, N. Decision-support models for empiric antibiotic selection in Gram-negative bloodstream infections. *Clin. Microbiol. Infect.* **2019**, *25*, 108.e1–108.e7. [CrossRef]
35. Lodise, T.P.; Bonine, N.G.; Ye, J.M.; Folse, H.J.; Gillard, P. Development of a bedside tool to predict the probability of drug-resistant pathogens among hospitalized adult patients with gram-negative infections. *BMC Infect. Dis.* **2019**, *19*, 718. [CrossRef] [PubMed]
36. Moons, K.G.; Altman, D.G.; Reitsma, J.B.; Ioannidis, J.P.; Macaskill, P.; Steyerberg, E.W.; Vickers, A.J.; Ransohoff, D.F.; Collins, G.S. Transparent Reporting of a multivariable prediction model for Individual Prognosis or Diagnosis (TRIPOD): Explanation and elaboration. *Ann. Intern. Med.* **2015**, *162*, W1–W73. [CrossRef]
37. Moons, K.G.M.; Wolff, R.F.; Riley, R.D.; Whiting, P.F.; Westwood, M.; Collins, G.S.; Reitsma, J.B.; Kleijnen, J.; Mallett, S. PROBAST: A Tool to Assess Risk of Bias and Applicability of Prediction Model Studies: Explanation and Elaboration. *Ann. Intern. Med.* **2019**, *170*, W1–W33. [CrossRef] [PubMed]
38. Steyerberg, E.W.; Vickers, A.J.; Cook, N.R.; Gerds, T.; Gonen, M.; Obuchowski, N.; Pencina, M.J.; Kattan, M.W. Assessing the performance of prediction models: A framework for traditional and novel measures. *Epidemiology* **2010**, *21*, 128–138. [CrossRef] [PubMed]
39. Moons, K.G.; Kengne, A.P.; Grobbee, D.E.; Royston, P.; Vergouwe, Y.; Altman, D.G.; Woodward, M. Risk prediction models: II. External validation, model updating, and impact assessment. *Heart* **2012**, *98*, 691–698. [CrossRef] [PubMed]
40. Snell, K.I.E.; Levis, B.; Damen, J.A.A.; Dhiman, P.; Debray, T.P.A.; Hooft, L.; Reitsma, J.B.; Moons, K.G.M.; Collins, G.S.; Riley, R.D. Transparent reporting of multivariable prediction models for individual prognosis or diagnosis: Checklist for systematic reviews and meta-analyses (TRIPOD-SRMA). *BMJ* **2023**, *381*, e073538. [CrossRef] [PubMed]
41. Moons, K.G.; de Groot, J.A.; Bouwmeester, W.; Vergouwe, Y.; Mallett, S.; Altman, D.G.; Reitsma, J.B.; Collins, G.S. Critical appraisal and data extraction for systematic reviews of prediction modelling studies: The CHARMS checklist. *PLoS Med.* **2014**, *11*, e1001744. [CrossRef]
42. Schardt, C.; Adams, M.B.; Owens, T.; Keitz, S.; Fontelo, P. Utilization of the PICO framework to improve searching PubMed for clinical questions. *BMC Med. Inform. Decis. Mak.* **2007**, *7*, 16. [CrossRef]
43. Damen, J.A.A.; Moons, K.G.M.; van Smeden, M.; Hooft, L. How to conduct a systematic review and meta-analysis of prognostic model studies. *Clin. Microbiol. Infect.* **2023**, *29*, 434–440. [CrossRef]
44. Viechtbauer, W. Conducting Meta-Analyses inRwith themetaforPackage. *J. Stat. Softw.* **2010**, *36*, 1–48. [CrossRef]

MDPI AG
Grosspeteranlage 5
4052 Basel
Switzerland
Tel.: +41 61 683 77 34

Antibiotics Editorial Office
E-mail: antibiotics@mdpi.com
www.mdpi.com/journal/antibiotics

www.ingramcontent.com/pod-product-compliance
Lightning Source LLC
Chambersburg PA
CBHW062058220926
51850CB00010B/2812
* 9 7 8 3 7 2 5 8 7 6 2 5 9 *